Berichte aus dem Institut für Mehrphasenströmungen

Band 6

Investigation and Modelling of Vortex Development and Gas Entrainment in Pump Intakes under Critical Inflow Conditions

Dem Promotionsausschuss der
Technischen Universität Hamburg

zur Erlangung des akademischen Grades
Doktor-Ingenieur (Dr.-Ing.)

vorgelegte Dissertation

von
Nicolai Sebastian Szeliga, M.Sc.

aus
Reinbek

2019

Bibliografische Information der Deutschen Nationalbibliothek
Die Deutsche Nationalbibliothek verzeichnet diese Publikation in der
Deutschen Nationalbibliografie; detaillierte bibliographische Daten sind im Internet
über http://dnb.d-nb.de abrufbar.
1. Aufl. - Göttingen: Cuvillier, 2020
Zugl.: (TU) Hamburg, Univ., Diss., 2020

1. Gutachter: Prof. Dr.-Ing. Michael Schlüter
2. Gutachter: Prof. Dr.-Ing. Dr. h.c. Uwe Hampel
Vorsitzender des Prüfungsausschusses: Prof. Dr.-Ing. Georg Fieg
Tag der mündlichen Prüfung: 24.02.2020

Nonnenstieg 8, 37075 Göttingen
Telefon: 0551-54724-0
Telefax: 0551-54724-21
www.cuvillier.de

1. Auflage, 2020
Gedruckt auf umweltfreundlichem, säurefreiem Papier aus nachhaltiger Forstwirtschaft.

ISBN 978-3-7369-7189-9
eISBN 978-3-7369-6189-0

Vorwort

Die vorliegende Arbeit entstand während meiner Zeit als wissenschaftlicher Mitarbeiter am Institut für Mehrphasenströmungen der Technischen Universität Hamburg unter der Leitung von Professor Dr.-Ing. Michael Schlüter, wo ich von November 2014 bis Dezember 2016 das Teilprojekt A des vom Bundesministerium für Bildung und Forschung (BMBF) geförderten Verbundprojektes SAVE[1] und von Januar 2017 bis März 2019 weitere Industrie- und Forschungsprojekte im Bereich der Mehrphasenströmungen bearbeitet habe. Grundlage dieser Dissertation ist dabei die im Rahmen des Verbundprojektes SAVE durchgeführten Untersuchungen. Für die finanzielle Förderung danke ich dem BMBF, für die Projektabwicklung möchte ich mich zudem bei Herrn Alexander Ehrlich vom Projektträger Karlsruhe (PTKA) bedanken.

Besonderer Dank gilt meinem Doktorvater Herrn Prof. Dr.-Ing. Michael Schlüter für die Betreuung dieser Arbeit, den interessanten Gesprächen und dass ich die Möglichkeit bekommen habe an vielen interessanten Projekten aus verschiedensten Bereichen mitzuarbeiten und auch selbstständig leiten zu dürfen. Gerade diese Vielseitigkeit an Forschungsthemen, in Verbindung mit tiefgreifenden fachlichen Diskussionen, bildet die Grundlage für den Erwerb eines breiten Erfahrungsschatzes, welcher mir in meiner beruflichen Laufbahn sicherlich noch von großem Nutzen sein wird.

Weiterhin möchte ich mich auch herzlich bei allen Projektpartnern für die anregenden Diskussionen und Projekttreffen bedanken. Besonderer Dank gilt insbesondere Herrn Prof. Dr.-Ing. Dr. h.c. Uwe Hampel für seine Bereitschaft meine Arbeit als Zweitgutacher zu bewerten und Herrn Dr. Frank Blömeling für seine vielen hilfreichen Ratschläge.

Mein Dank geht auch an Herrn Prof. Dr.-Ing- Georg Fieg für die Übernahme des Prüfungsvorsitzes.

Bei meinen Kollegen vom Institut, von denen ich zu meiner großen Freude auch viele als Freunde zählen kann, möchte ich mich ebenfalls besonders herzlich bedanken. An die

[1] Verbundprojekt SAVE: *„**S**icherheitsrelevante **A**nalyse des **V**erhaltens von Armaturen, Kreiselpumpen und **E**inlaufgeometrien unter Berücksichtigung störfallbedingter Belastungen"* (FKZ: 02NUK023A)

große Hilfsbereitschaft untereinander, die spannenden Unterhaltungen in der Kaffeepause, die gemeinsamen Konferenzen, Fußballspiele und Treffen nach Feierabend werde ich mich immer gerne zurückerinnern. Besonderer Dank geht an Herrn Steffen Richter und Herrn Christian Busch für den Aufbau der Versuchsanlage und an Herrn Daniel Bezecny für die Durchführung von numerischen Wirbelsimultionen.

Ich danke auch den Sekretärinnen und Technikern des Instituts, der Forschungswerkstatt unter der Leitung von Herrn Dirk Manning und meinen Studenten, die mich alle sehr bei meiner Arbeit unterstützt haben. Namentlich erwähnen möchte ich Herrn Felix Kexel, der mich über fast meine gesamte Zeit am Institut als studentische Hilfskraft unterstützt hat und Herrn Lando Helmrich von Elgott, der mit seiner hervorragenden Bachelorarbeit wichtige Erkenntnisse beigesteuert hat.

Großer Dank geht auch an meine Freunde und meine Familie, insbesondere an meine Frau Elena und meinen Sohn Sascha, die mir immer zur Seite standen und in stressigen Zeiten für Ausgleich sorgten.

Ewiger Dank gilt meinen Eltern, Angelika und Bernd, denen ich diese Arbeit widmen möchte. Ohne ihre bedingungslose Liebe und Unterstützung wäre ich niemals so weit gekommen.

Frankfurt am Main, im März 2020

Contents

Figures

Tables

Notations

Latin Symbols:

Symbol:	Name:	Unit:
A	Area	m^2
a	Downward acceleration	s^{-1}
B	Channel width	m
b	Thickness	m
C	Circle line	-
C	Fitting parameter	m^{-1}
c	Spin constant	$m^2 \cdot s^{-1}$
D	Diameter	m
d	Diameter	m
f	Recording frequency	Hz
g	Gravitational acceleration	$m \cdot s^{-2}$
H	Height	m
h	Height	m
K	Viscosity function	-
k	Pump intake orientation factor	-
L	Gas-core length	m
p	Pressure	Pa
Q	Volume flow rate	$m^3 \cdot s$
r	Radius	m
S	Submergence depth	m
T	Temperature	°C

t	Time	s
u	Velocity	$m \cdot s^{-1}$
V	Volume	m^3
z	Axial coordinate	m

Greek Symbols:

Symbol:	Name:	Unit:
α	Approaching flow angle	°
α	Fitting exponent	-
β	Vertical inflow angle	°
Γ	Circulation	$m^2 \cdot s^{-1}$
λ	Scaling factor	-
λ	Wave length	nm
μ	Dynamic viscosity	Pa·s
ν	Kinematic viscosity	$m^2 \cdot s^{-1}$
ρ	Density	$kg \cdot m^{-3}$
σ	Surface tension	$N \cdot m^{-1}$
φ	Pump intake orientation	-
Ψ	Stream function	$m^4 \cdot s^{-1}$

Dimensionless numbers:

Symbol:	Name:	Equation:
$\boldsymbol{C_{spin}}$	Spin parameter	$C_{\mathrm{spin}} = \frac{c}{\sqrt{g \cdot D^3}}$
$\boldsymbol{L^*}$	Dimensionless gas-core length	$L^* = \frac{L}{S}$
$\boldsymbol{Re_r}$	Radial Reynolds number	$Re_{\mathrm{r}} = \frac{Q_{\mathrm{r}}}{\nu}$
$\boldsymbol{S^*}$	Dimensionless submergence	$S^* = \frac{S}{D}$
$\boldsymbol{\bar{\Gamma}}$	Dimensionless circulation (Granger model)	$\bar{\Gamma} = \frac{\Gamma}{\Gamma_\infty}$
$\boldsymbol{\Gamma^*}$	Dimensionless circulation	$\Gamma^* = \frac{\Gamma \cdot D}{Q}$
$\boldsymbol{\Gamma_S^*}$	Dimensionless circulation (Jain correlation)	$\Gamma_{\mathrm{S}}^* = \frac{\Gamma \cdot S}{Q}$
$\boldsymbol{\bar{\Psi}}$	Dimensionless stream function	$\bar{\Psi} = \frac{\Psi}{Q_{\mathrm{r}} \cdot h}$
$\boldsymbol{\nu^*}$	Dimensionless viscosity	$\nu^* = \frac{\sqrt{g \cdot D^3}}{\nu}$
$\boldsymbol{Fr}$	Froude number	$Fr = \frac{u}{\sqrt{g \cdot D}}$
$\boldsymbol{Re}$	Reynolds number	$Re = \frac{u \cdot D}{\nu}$
$\boldsymbol{We}$	Weber number	$We = u \cdot \sqrt{\frac{\rho \cdot D}{\sigma}}$
$\boldsymbol{\eta}$	Dimensionless radial component	$\eta = \left(\frac{r}{r_0}\right)^2$
$\boldsymbol{\xi}$	Dimensionless axial component	$\xi = \frac{z}{h}$

Indices:

Symbol:	Name:
*	Normalized/dimensionless
0	Vortex-core
∞	Bulk phase
ANSI	Result based on ANSI correlation
BR	Result based on Burgers-Rott model
CFD	Numerical result
crit	Critical value
crit,90	90 % of the critical value
exp	Experimental result
floor	Intermediate floor
Intake	Pump intake opening
LDV	Laser Doppler Velocimetry value
LP	Laboratory plant setup
min	Minimal value
p	Seeding particles
PP	Pilot plant setup
r	Radial
R	Reservoir
S	Submergence depth based
z	Axial
θ	Azimuthal
ν	Viscous

Abbreviations:

Symbol:	Name:
CAD	Computer-Aided Design
CFD	Computational Fluid Dynamics (numerical simulations)
DI	De-Ionized (water)
DLSR	Digital single-lens reflex camera
DN	Nominal Diameter
LDV	Laser Doppler Velocimetry
PIV	Particle Image Velocimetry
PVC-U	Unplasticized Polyvinyl chloride
TUHH	Hamburg University of Technology

Zusammenfassung

Gaseintrag durch Ausbildung von freien Oberflächenwirbeln in Pumpenansaugbecken ist ein ernstes, aber oftmals vernachlässigtes, Sicherheitsrisiko in vielen industriellen Prozessen. Aufgrund der komplexen strömungsmechanischen Phänomene, die bei der Wirbelbildung auftreten sowie die große Anzahl an Ansauggeometrien und –größen, sind die korrekte Vorhersage der Wirbelentstehung und deren Einfluss auf ein System sehr herausfordernd.
Bestehende Wirbelmodelle beschränken sich oft auf eine vereinfachte theoretische Beschreibung der Wirbelbildung oder Experimente im Labormaßstab, da die benötigten Modellparameter nur schwer oder nach dem heutigen Stand der Technik unmöglich im größeren Maßstab zu messen sind. Auf der anderen Seite sind die Annahmen für die Korrelationen zur kritischen Überdeckung oftmals zu stark vereinfachend, insbesondere in Systemen mit hoher Zirkulation, wodurch die kritische Überdeckung überschätzt wird.

Damit die Genauigkeit der Korrelation und die Anwendbarkeit der Wirbelmodelle verbessert und der Einfluss von starker Zirkulation auf die Wirbelbildung untersucht wird, sind zwei experimentelle Versuchsanlagen an der Technischen Universität Hamburg (TUHH) im Rahmen des Verbundprojektes SAVE errichtet worden. Beide Versuchsanlagen bestehen dabei aus je einem zylindrischen Pumpenansaugbecken mit vertikal-abwärts gerichtetem Pumpenzulauf. Die erste Anlage ist eine Pilotanlage im Technikumsmaßstab, mit einem Beckenvolumen von 50 m^3 und einem nominellen Rohrdurchmesser von DN200. Die zweite Anlage ist geometrisch ähnlich, jedoch um den Faktor 13,3 nach unten skaliert. Diese Laboranlage besitzt ein Reservoir mit einem Volumen von 20 L und einem nominellen Rohrdurchmesser von DN15. In beiden Anlagen wird Wasser mittels einer Kreiselpumpe aus dem Becken gefördert und über eine geschlossene Rohrleitung von oben wieder zurück ins Becken geleitet. Die Zuleitung besteht dabei aus vier über dem Umfang angeordnete Rohrleitungen, welche das Wasser in einem definierten Winkel zurück in das Becken leiten und dadurch gezielt eine Zirkulation aufprägen.

Die durchgeführten Experimente beinhalten die Messung von Luftkernlängen, Wirbelstärke und Strömungsfeldern unter Variation des Förderstroms, der Überdeckungshöhe, des Einlaufwinkels, der Ansaugöffnung sowie der Verwendung von Wirbelbrechern. Die Messungen des Strömungsfeldes erfolgen dabei mittels Particle Image Velocimetry (PIV).

Die Ergebnisse zeigen unter Anderem, dass eine hohe Zirkulation nicht nur die Wirbelstärke, sondern auch die Stabilität und damit die Vorhersagbarkeit erhöht. Hohe Zirkulation reduziert weiterhin den Einfluss der Viskosität und der Oberflächenspannung auf die Ausformung eines Luftkerns, insbesondere im Labormaßstab, wo diese Effekte einen größeren Einfluss haben. Aus den gemessenen Geschwindigkeitsprofilen und unter Anwendung des von *Ito et al.* verbesserten Wirbelmodells von *Burgers & Rott*, können die sich ausbildenden Luftkernlängen korrekt vorhergesagt werden.

Die Ergebnisse zeigen außerdem einen Zusammenhang zwischen dem Wirbelkernradius und dem Durchmesser der Ansaugleitung wodurch die Anwendbarkeit des Wirbelmodells verbessert werden kann. Mit den Wirbelbrechern ist eine einfache und kostengünstige Methode evaluiert worden, welche Gaseintrag in die Pumpe selbst bei hoher Zirkulation und großen Förderströmen effektiv und zuverlässig verhindert.

Abstract

Gas entrainment due to the formation of free surface vortices is a serious, but often neglected, safety hazard in many industrial processes. The prediction of vortex development and its influence on a system are difficult to estimate, due to the complexity of the underlying fluid mechanics, and the huge variety of scales and geometries in which vortices can occur.
Existing vortex models are mostly limited to theoretical observations and laboratory experiments, since the variables needed to calculate the gas-core formation are hard or even downright impossible to measure in a real pump intake structure. Correlations for the critical submergence depth on the other hand, offer only rough estimations with high uncertainties, especially in systems with a high circulation.

To improve the predictability and to quantify the influence of induced circulation on the vortex formation, two experimental setups have been built at the Hamburg University of Technology (TUHH) in the scope of the project SAVE. Both setups are geometrically similar and consist of cylindrical shaped reservoirs with vertical pump intakes. The first one is of pilot plant size with a reservoir volume of 50 m^3 and a pipe diameter of DN200. The second one is of laboratory scale with a reservoir volume of 20 L and a pipe diameter of DN15. In both cases the water is pumped in a loop from the pump intake back into the reservoir through four circularly arranged entry pipes with defined inflow angles to induce varying strengths of circulation. Experiments have been conducted to measure the influence of intake geometries, vortex breakers, submergence depths, inflow angles and volume flow rates on the vortex strength and the length of the occurring gas cores. Additionally, Particle Image Velocimetry (PIV) measurements and coloring experiments with ink have been carried out, to characterize the velocity fields and the velocity profiles inside the vortices.

Results not only show how high circulation values increase the gas-core length, but in fact stabilize the vortex. A strong circulation also negates the influences of viscosity and surface tension, especially in small scale systems, where their influence is higher. The measured velocity profiles have been used to correctly predict the occurring gas-core lengths with help of the vortex model by *Burgers & Rott* with the improvements made by *Ito et al.* A correlation between the vortex-core radius and the intake radius has been found, which further simplifies the model and increases its practical usability. At last, simple vortex suppressing measures have proved to offer a simple, low cost solution to drastically hinder the vortex formation and therefore, reduce gas entrainment into the pump system.

1 Introduction

The occurrence of gas-entraining free surface vortices at pump intakes poses a huge safety hazard for the reliable operation of cooling circuits in chemical and power plants. Without proper safety precautions, entrained gas will accumulate inside the system and cause the breakdown of coolant flow as well as long-term damage through cavitation inside pumps, valves and heat exchangers.

Furthermore, free surface vortices are the cause of problems in many other fields of operation, since they can occur anywhere, where liquids are drained from reservoirs, e.g. the sump of a rectification column, storage tanks for chemical processes and even at hydro-electric power turbines.

Vortex formation is a complex topic, depending on many different parameters, like reservoir size and shape, volume flow rate, liquid level, induced momentum and fluid properties. The phenomena of vortex development, the fluid mechanics behind it and the existing theoretical vortex models and safety correlations will therefore be discussed in detail in chapter 2. Numerous groups from a diverse range of scientific and engineering fields have conducted research, both experimental and theoretical, on this topic with different approaches and varying degrees of success. Through their culminated effort a plethora of vortex models and correlations has been developed over the last decades, of which none can be applied universally so far. Most models and correlations are either too oversimplified, limited in their use for certain applications or too complex to be useful in practice.

Advancements in the field of optical measurement techniques as well as increased computational power in the recent decade enable a more sophisticated approach on the experimental and numerical examination of vortex formation. With the help of the thereby gained research data the existing models and correlations can be evaluated and, if deemed useful, enhanced to provide more accurate predictions of the vortex development and the risk of gas entrainment.

1.1 Research Goals

The work conducted within this thesis is performed as part of the project SAVE: "***S****afety-relevant* ***a****nalysis of the performance of centrifugal pumps,* ***v****alves and inlet geometries, including stress-related* ***e****vents*" (*German*: Verbundprojekt SAVE: „***S****icherheitsrelevante* ***A****nalyse des* ***V****erhaltens von Armaturen, Kreiselpumpen und* ***E****inlaufgeometrien unter Berücksichtigung störfallbedingter Belastungen*", German Federal Ministry of Education and Research (BMBF), grant number 02NUK023A). The research goals of this thesis are to evaluate the existing vortex models and correlations and to improve suitable models towards a more practical usability. Further, design criteria are investigated and ranked, based on their impact on the vortex development and recommendations for the design of pump intakes are developed.

For the experimental investigation of the vortex formation, a large-scale pilot plant is constructed in the experimental hall of the Hamburg University of Technology (TUHH) as well as a smaller, geometrical similar laboratory setup. Experiments include the measurement of gas-core lengths, velocity fields and profiles of free surface vortices. Investigated variables besides the scale are: Volume flow, liquid level, pump intake size and shape and induced circulation. The experimental setup as well as the conducted experiments are described in detail in chapter 3, the results and discussion can be found in chapter 4.

2 Theoretical Background

In this chapter the phenomenon of vortex formation is described in detail, from the fundamentals of vortex development and fluid dynamics up to vortex scaling criteria and vortex preventing devices. A focus is set on prior research conducted by other research groups, including their vortex models and correlations for critical submergence depths, emphasizing on models and correlation which are applicable for the experimental setup used in this thesis.

2.1 Fundamentals of Vortex Development

In fluid dynamics, a vortex is a coherent structure within a moving fluid, with a strong rotational flow. Intake vortices are a specific form of vortices forming around an axial downdraft, often caused by the pressure drop of an outflow of liquid from a reservoir. At first the liquid is flowing towards the downdraft in radial direction, but due to asymmetries in the flow, the liquid starts to rotate with increasing azimuthal velocity towards the center of the downdraft, thus forming an intake vortex [Nog03]. The main causes for asymmetric flow are design based, like an eccentric flow around a corner or obstacle as well as velocity gradients in the approaching flow, due to viscous friction on the reservoir walls, see Figure 2-1 for a graphic display of the main causes [Hec78].

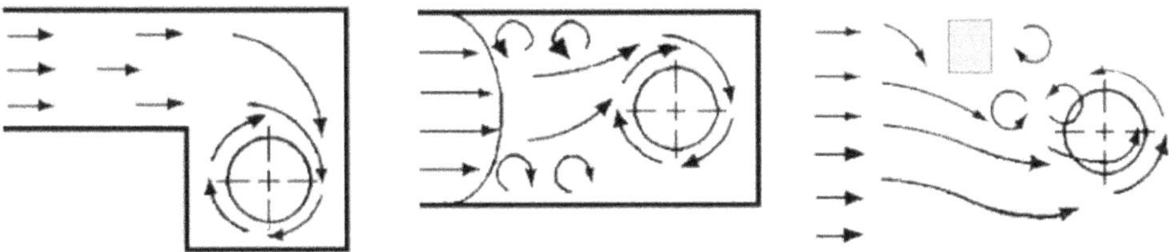

Figure 2-1: Sources of flow circulation (From left to right: (i) eccentricity, (ii) velocity gradient, and (iii) obstruction (adapted from *Hecker & Durgin,* 1978) [Auc09].

Vortices can be classified in three different ways according to *Knauss*: The point of origin, the shape and the duration of occurrence. Rotation free vortices, also called concentric vortices, occur on resting liquid surfaces when the submergence depth is small in proportion to the intake opening size. Rotating vortices can occur either on free liquid surfaces or as submerged vortices on walls and floors within the liquid body, see Figure 2-2 and Figure 2-3 for a schematic visualization of the vortex types.

Submerged vortices do not cause gas entrainment but may introduce angular momentum into the pump intake. Rotating vortices on the liquid surface, also called free surface vortices, can cause both momentum and gas entrainment into the pump intake [Kna83]. Furthermore, all forms of vortices can be divided into steady and unsteady vortices [Gül13].

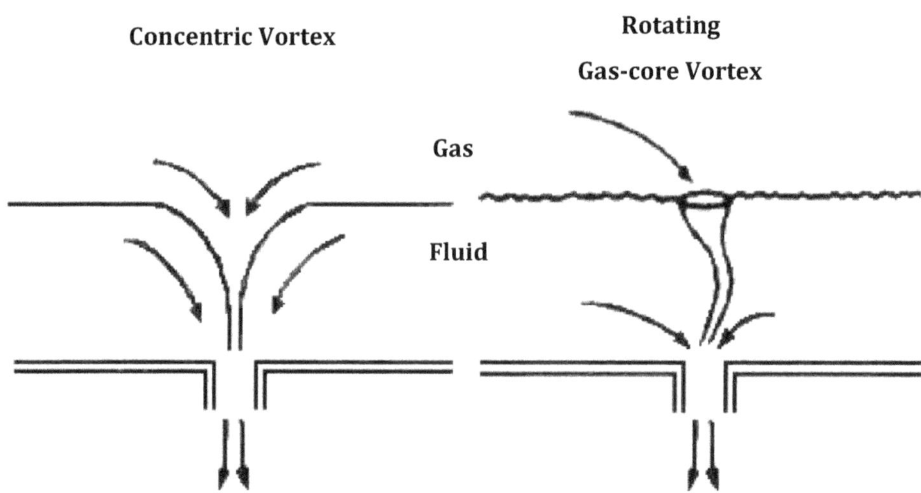

Figure 2-2: Schematic visualization of surface verticesaccording to *Knauss* [Rei82].

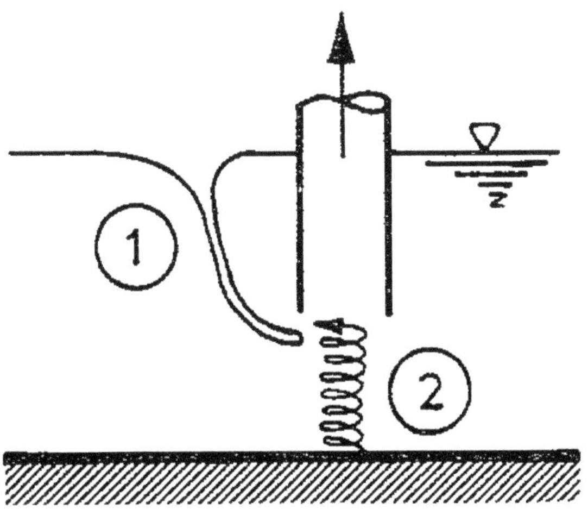

Figure 2-3: Schematic visualization of rotating vortex types: (1) Free surface vortex, (2) submerged vortex [Kna83].

Free rotating surface vortices can be classified further by their strength, most visible by the length of the formed gas core, see Figure 2-4. The vortices can be separated into six different strength levels, based on their gas-core lengths. The lowest vortex strength level is 1, where neither a gas-core nor rotation on the surface are visible yet and only subsurface rotation occurs. Level 2 sees the start of surface rotation as well as the formation of a surface dimple, while at level 3 a fully developed gas-core is visible. Level 3 also marks the development stage, where the invisible vortex core first reaches into the pump intake, which can be visualized by dye-core experiments. Vortices of strength level 4 have a longer gas-core and particles (if present in the fluid) start to accumulate at the gas-core tip. Gas-entrainment into the pump systems start at vortex strength level 5, where gas-bubbles separate from the gas-core tip. Level 6, marks critical vortex conditions, where the gas-core reaches into the pump intake and gas is entrained continuously. Vortices of the levels 1 to 4 are generally tolerable for the operation of a pump system since they only introduce momentum to the system. Vortices of level 3 and 4 should still be avoided if possible, since they can lead to the entrainment of solid objects into the system, which can clog pipes or damage pumps. Vortices of level 5 and 6 are to be avoided, as they are the cause for gas entrainment into the system, which can accumulate in the pump impeller and cause cavitation damage and the collapse of the flow rate [Hec78]. See Figure 2-5 for an example of a level 6 vortex, recorded in the pilot plant setup.

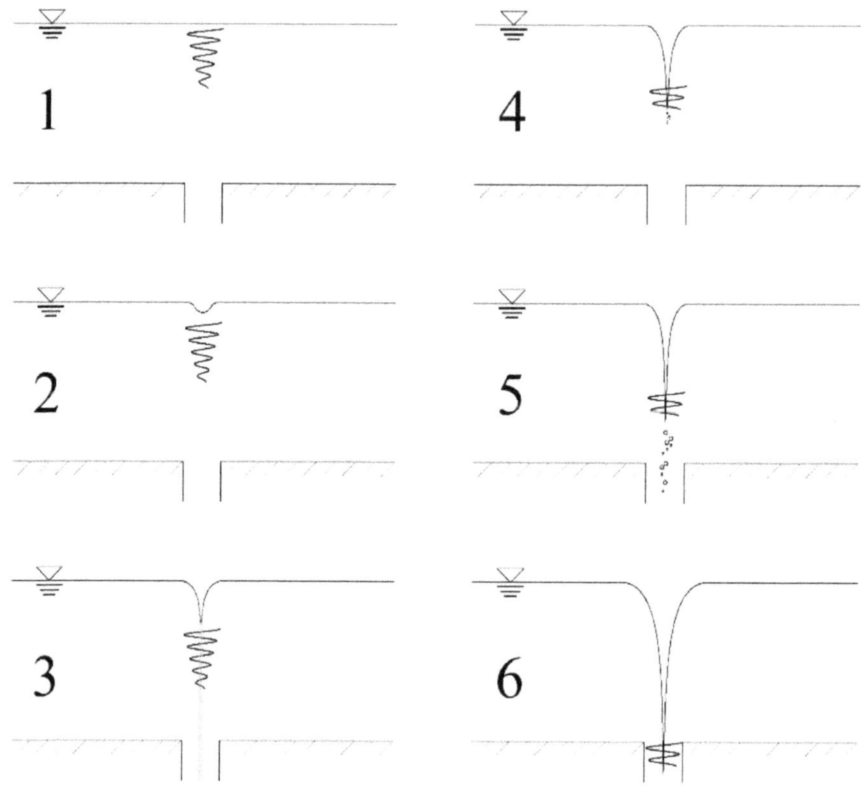

Figure 2-4: Classification of vortex strengths according to *Hecker & Durgin*. 1. Subsurface rotation, 2. Forming of first surface dimple, 3. Vortex-core reaches into the pump intake, 4. Entrainment of particles at the gas-core tip, 5. Start of gas (bubble) entrainment into the pump intake, 6. Critical vortex condition – Gas-core reaches into the pump intake [Hec78].

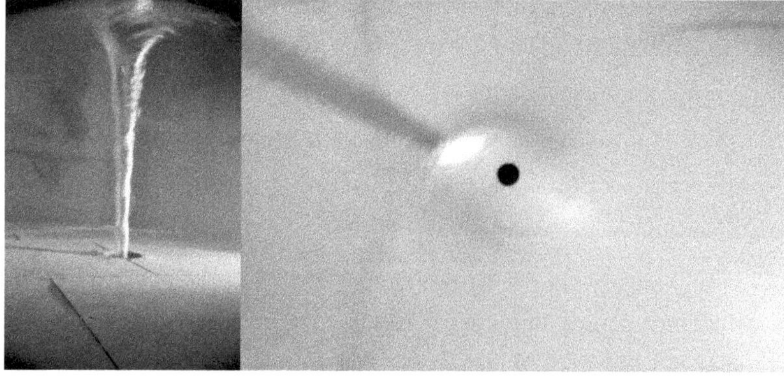

Figure 2-5: Photograph of a level 6 vortex with fully developed gas-core. Recorded in the pilot plant setup. Left: side view, right: Top view.

2.1.1 Parameters and Dimensionless Numbers

To be able to correctly describe the vortex development, either with theoretical models (see subchapter 2.2) or empirical correlations (see subchapter 2.3), the relevant parameters for the vortex development must be identified. The parameters can be divided into fluid dynamic properties and geometrical parameters. Fluid properties are:

- Density ρ,
- Viscosity - kinematic ν or dynamic μ,
- Surface tension σ,
- Circulation Γ (a measure for the vortex strength, see p. 9),
- Submergence depth S (the height of the water level above the pump intake),
- Intake velocity u and volume flow rate Q.

The relevant geometrical parameters are the intake shape and intake pipe diameter d, the orientation of the intake (horizontal, vertical upwards or downwards, traverse), the size and shape of the intake chamber, whether the intake is connected to the wall or protrudes into the chamber and whether vortex breakers (see subchapter 2.5) are used [Jai78, Kna83].

An important parameter to determine the influence of vortex development on a pump system is the critical submergence depth S_{crit}. The critical submergence depth is defined as the depth, for which a gas-core vortex would reach the pump intake [Jai78].

By combining the correct parameters, dimensionless number can be used to compare different designs and sizes. Important dimensionless numbers are the Froude number

$$Fr = \frac{u}{\sqrt{g \cdot D}} \qquad (2\text{-}1),$$

which describes the inertial to weight force ratio of a fluid system, with g as the gravitational acceleration, the Reynolds number

$$Re = \frac{u \cdot D}{\nu} \qquad (2\text{-}2),$$

describing the ratio of inertial to viscous forces, and the Weber number

$$We = u \cdot \sqrt{\frac{\rho \cdot D}{\sigma}} \qquad (2\text{-}3),$$

which defines the ratio of inertial to surface tension forces. Further, the dimensionless circulation

$$\Gamma^* = \frac{\Gamma \cdot D}{Q} \qquad (2\text{-}4),$$

also known as the circulation number, and the dimensionless submergence

$$S^* = \frac{S}{D} \qquad (2\text{-}5),$$

are used for the description of vortices in intake systems [Kna83].

In this thesis, another dimensionless quantity is introduced. Dividing the gas-core length L by the submergence depth S, yields the dimensionless gas-core length

$$L^* = \frac{L}{S} \qquad (2\text{-}6),$$

with values between 0, where no gas-core has formed yet, and 1, where the gas-core reaches the pump intake and $S = S_{crit}$. The dimensionless gas-core length is used to compare the different experimental sizes and submergence depths with each other. [Sze16]

2.2 Theoretical Vortex Models

The calculation of the vortex fluid dynamics in pump intakes and the correct prediction of the occurring gas core lengths and amount of entrained gas is challenging, due to the complex three-dimensional flow and the vast amount of intake geometries and sizes. There have been many attempts to develop a unified vortex model for all possible cases, but so far, no universally applicable model has been found. In the following chapters the most widely used vortex models are described and their strengths and shortcomings are explained. While most models are shorty summarized, the model of *Burgers & Rott*, used for the evaluation, is explained in detail, see chapter 2.2.3.

2.2.1 Rankine Model

This mathematical vortex model is the first vortex model, proposed in 1858 by *William. J. M. Rankine*. It is the basis for many following vortex models and still in use today, due to its simplicity, i.e. *Keller et al.* [Kel14]. According to *Rankine*, the vortex can be divided into two regions. The vortex core region, which behaves like a rigid body rotation and the outer vortex, which behaves like an irrotational vortex. For the regions, *Rankine* defined two separate equations for the azimuthal velocity

$$u_\theta = \begin{cases} \frac{\Gamma_\infty}{2\pi} \cdot \frac{r}{r_0^2} & for\ r \leq r_0 \\ \frac{\Gamma_\infty}{2\pi r} & for\ r > r_0 \end{cases} \qquad (2\text{-}7)$$

over the radius r. With r_0 as the radius of the vortex core and Γ_∞ as the circulation outside the vortex region, also called the bulk circulation. Since the azimuthal velocity and its decrease is slow in the outer regions of the vortex (bulk phase), the bulk circulation can be seen as a constant value independent of the vortex radius r. The azimuthal velocity profile over the radius can be seen in Figure 2-6. Inside the core region, the azimuthal velocity is raising linear, until the border of the vortex core is reached. Outside the core region the velocity is then decreasing hyperbolically.

The bulk circulation is calculated by integrating the azimuthal velocity over a circle line C

$$\Gamma_\infty = \oint_C u_\infty \cdot dC \qquad (2\text{-}8),$$

where u_∞ is the velocity vector and dC the azimuthal circle line [Ran58]. See Figure 2-7 for an exemplified view.

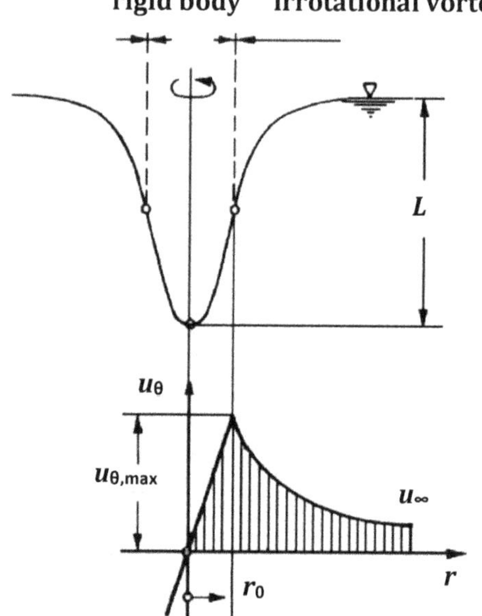

Figure 2-6: Schematic view of the Rankine vortex model with the course of the azimuthal velocity profile over the radius and the profile of the liquid surface above, with L as the resulting gas core length [Kna83].

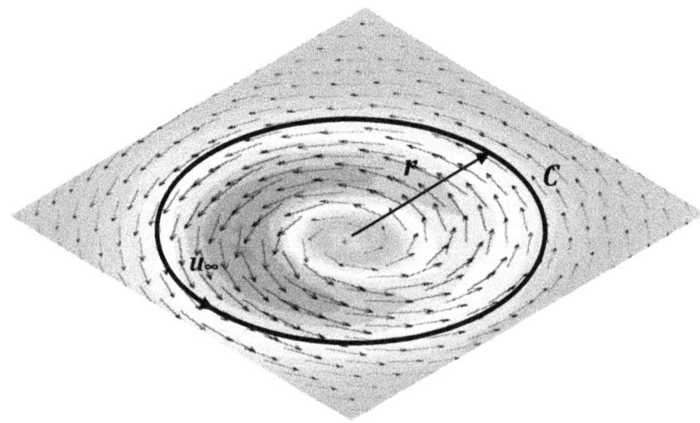

Figure 2-7: 2D-Particle Image Velocimetry measurement plot of a free surface vortex, with the circle line C, the radius r and the velocity along the circle line u.

2.2.2 Lamb–Oseen vortex

A Lamb-Oseen vortex is an exact solution of the Navier-Stokes equations (see eq. 2-11 to 2-13) for a viscosity driven decay of a potential vortex. The azimuthal velocity can be expressed by

$$u_\theta(r,t) = \frac{\Gamma_\infty}{2\pi r} \cdot \left[1 - exp\left(-\frac{r^2}{r_0^2(t)}\right)\right] \quad \text{(2-9)},$$

where $r_0(t) = \sqrt{4\nu t}$ is the time depended vortex core radius, with the kinematic viscosity ν and the time t [Wu06]. Since the vortices observed in this thesis are stationary, the Lamb-Ossen vortex model is not further regarded.

2.2.3 Burgers & Rott

The vortex model of *Burgers & Rott*, also simply known as *Burgers vortex*, is a mathematical approach, based on hydrodynamic principles, first developed by *Johannes M. Burgers* [Bur48] and later adjusted by *Nicolas Rott* [Rot58]. The model is derived for a single phase inside a cylindrical system from the continuity equation

$$\frac{\partial \rho}{\partial t} + \frac{1}{r}\frac{\partial}{\partial r}(r\rho u_r) + \frac{1}{r}\frac{\partial}{\partial \theta}(\rho u_\theta) + \frac{\partial}{\partial z}(\rho u_z) = 0 \quad \text{(2-10)},$$

and the Navier-Stokes equations in radial

$$\begin{aligned} &\rho\left(\frac{\partial u_r}{\partial t} + u_r\frac{\partial u_r}{\partial r} + \frac{u_\theta}{r}\frac{\partial u_r}{\partial \theta} + u_z\frac{\partial u_r}{\partial z} - \frac{u_\theta^2}{r}\right) = \\ &-\frac{\partial p}{\partial r} + \mu\left(\frac{\partial^2 u_r}{\partial r^2} + \frac{1}{r}\frac{\partial u_r}{\partial r} + \frac{1}{r^2}\frac{\partial^2 u_r}{\partial \theta^2} + \frac{\partial^2 u_r}{\partial z^2} - \frac{u_r}{r^2} - \frac{2}{r^2}\frac{\partial u_\theta}{\partial \theta}\right) + \rho g_r \end{aligned} \quad \text{(2-11)},$$

azimuthal

$$\begin{aligned} &\rho\left(\frac{\partial u_\theta}{\partial t} + u_r\frac{\partial u_\theta}{\partial r} + \frac{u_\theta}{r}\frac{\partial u_\theta}{\partial \theta} + u_z\frac{\partial u_\theta}{\partial z} - \frac{u_r u_\theta}{r}\right) = \\ &-\frac{1}{r}\frac{\partial p}{\partial \theta} + \mu\left(\frac{\partial^2 u_\theta}{\partial r^2} + \frac{1}{r}\frac{\partial u_\theta}{\partial r} + \frac{1}{r^2}\frac{\partial^2 u_\theta}{\partial \theta^2} + \frac{\partial^2 u_\theta}{\partial z^2} + \frac{2}{r^2}\frac{\partial u_r}{\partial \theta} - \frac{u_\theta}{r^2}\right) + \rho g_\theta \end{aligned} \quad \text{(2-12)},$$

and axial direction

$$\begin{aligned} &\rho\left(\frac{\partial u_z}{\partial t} + u_r\frac{\partial u_z}{\partial r} + \frac{u_\theta}{r}\frac{\partial u_z}{\partial \theta} + u_z\frac{\partial u_z}{\partial z}\right) = \\ &-\frac{\partial p}{\partial z} + \mu\left(\frac{\partial^2 u_z}{\partial r^2} + \frac{1}{r}\frac{\partial u_z}{\partial r} + \frac{1}{r^2}\frac{\partial^2 u_z}{\partial \theta^2} + \frac{\partial^2 u_z}{\partial z^2}\right) + \rho g_z \end{aligned} \quad \text{(2-13)},$$

with density ρ, dynamic viscosity μ, pressure p, gravitational acceleration g, radial, azimuthal and axial velocity $u_\mathrm{r}, u_\theta, u_\mathrm{z}$, in the respective directions r, θ, z. The assumptions made for the model are that the system is stationary, axis-symmetrical and that the fluid phase is incompressible and inviscid. Using these assumptions, the continuity equation (eq. 2-10) can be simplified to

$$\frac{1}{r}\frac{\partial}{\partial r}(r\rho u_\mathrm{r}) + \frac{\partial}{\partial z}(\rho u_\mathrm{z}) = 0 \qquad (2\text{-}14)$$

and the Navier-Stokes equation in azimuthal direction (eq.2-12) to

$$u_\mathrm{r}\frac{\partial u_\theta}{\partial r} + u_\mathrm{z}\frac{\partial u_\theta}{\partial z} - \frac{u_\mathrm{r} u_\theta}{r} = \nu\left(\frac{\partial^2 u_\theta}{\partial r^2} + \frac{1}{r}\frac{\partial u_\theta}{\partial r} + \frac{\partial^2 u_\theta}{\partial z^2} - \frac{u_\theta}{r^2}\right) \qquad (2\text{-}15).$$

In the next step a potential flow is superimposed over a stagnation flow towards a vertical, round outlet on the bottom of the cylinder. The azimuthal velocity can therefore be expressed by

$$u_\theta(r) = \frac{\Gamma_\infty}{2\pi r} \cdot f(r) \qquad (2\text{-}16).$$

Inserting eq. 2-16 in eq. 2-15 and assuming height independency, yields

$$u_\mathrm{r}\frac{\Gamma_\infty}{2\pi}\left(\frac{\partial f(r)}{r \cdot \partial r} - \frac{f(r)}{r^2}\right) = \nu\frac{\Gamma_\infty}{2\pi}\left(\frac{\partial^2 f(r)}{r \cdot \partial r^2} + \frac{1}{r^2}\frac{\partial f(r)}{\partial r} - \frac{f(r)}{r^3}\right) \qquad (2\text{-}17).$$

Solving the partial derivative leads to

$$u_\mathrm{r}\frac{\Gamma_\infty}{2\pi}\left(\frac{f'(r)}{r} + \frac{f(r)}{r^2} - \frac{f(r)}{r^2}\right) = \nu\frac{\Gamma_\infty}{2\pi}\left(\frac{d}{dr}\left(\frac{f'(r)}{r}\right) - \frac{f'(r)}{r^2} + \frac{2 \cdot f(r)}{r^3} + \frac{f'(r)}{r^2} - \frac{2 \cdot f(r)}{r^3}\right) \qquad (2\text{-}18)$$

and crossing out the negating terms further leads to

$$u_\mathrm{r}\frac{f'(r)}{r} - \nu\frac{d}{dr}\left(\frac{f'(r)}{r}\right) = 0 \qquad (2\text{-}19),$$

with $f'(r) = df(r)/dr$. Inside the vortex core region, the radial velocity is assumed as

$$u_\mathrm{r} = -ar \quad (r < r_0) \qquad (2\text{-}20),$$

with the downward acceleration

$$a = \frac{2\nu}{r_\mathrm{v}^2} \qquad (2\text{-}21).$$

The downward acceleration, also called 'inflow gradient', is derived from the kinematic viscosity ν and the viscous radius r_ν of the vortex, which is often equaled with the vortex core radius ($r_\nu \approx r_0$) and assumed as height independent, but actually depending on z [Rot58]. *Odgaard* proposed a linear profile for the downward acceleration

$$a = \frac{\partial u_z}{\partial z} = \frac{u_z(z = S)}{S} \qquad (2\text{-}22),$$

from the fluid surface to the pump intake, with S as the submergence depth and $u_z(z = S)$ as the intake velocity [Odg86].

The axial velocity u_z can be calculated by inserting the radial velocity from eq. 2-20 into the modified continuous equation (eq. 2-14)

$$-\frac{a}{r}\frac{dr^2}{dr} + \frac{du_z}{dz} = 0 \qquad (2\text{-}23).$$

Solving the differentials and using Taylor's theorem, stopping after the first order, yields the equation for the axial velocity

$$u_z = 2az \qquad (2\text{-}24),$$

which is only linear depending on the axial coordinate z and independent of the radius r.

Now the only velocity missing is the azimuthal velocity u_θ, which can be calculated from eq. 2-19, with the radial velocity taken again from eq. 2-20 and the assumptions

$$\lim_{r \to \infty} f(r) = 1 \qquad (2\text{-}25)$$

and

$$lim_{r \to 0}\, u_\theta(r) = u_\theta(0) = 0 \qquad (2\text{-}26),$$

preventing a singularity for u_θ at the vortex center ($r = 0$). Solving the resulting ordinary differential equation yields the equation for the azimuthal velocity

$$u_\theta(r) = \frac{\Gamma_\infty}{2\pi r} \cdot \left[1 - exp\left(-\frac{ar^2}{2\nu}\right)\right] \qquad (2\text{-}27).$$

The equations for the radial and the axial velocity are simply derived from the stagnation flow. The radial velocity (eq. 2-20) is linear dependent on the radius and therefore tends to negative infinity when r goes to infinity. The axial velocity (eq. 2-24) is completely independent of r and instead linear depending of z Due to this, both equations are limited to the vortex core region ($r < r_0$).

The equation for the azimuthal velocity (eq. 2-27) is a continuous equation along the radius for $r > 0$, unlike the azimuthal velocity in the Rankine model (eq. 2-7) which possesses a discontinuity at $r = r_0$. Therefore, the Burgers-Rott vortex model offers a more realistic course of the azimuthal velocity profile, with a maximum at the viscous core radius r^*. The comparison of the two velocity profiles can be seen in Figure 2-8.

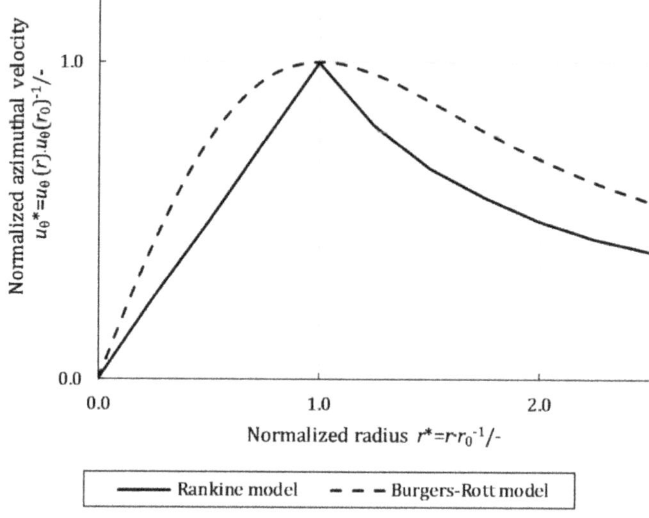

Figure 2-8: Comparison of the normalized azimuthal velocity profiles along the normalized radius between the Rankine and the Burgers-Rott vortex models.

The physical interpretation for the steady solution is given by *N. Rott*: *"While the vortex tends to decay, the 'onrushing' stagnation point flow carries new circulation from infinity towards the axis" (N. Rott: On the Viscous Core of a Line Vortex, Zeitschrift für angewandte Mathematik und Physik, ZAMP (1958), No. 9, p. 545)* [Rot58].

The model has been further improved by *Ito et al.* with the addition of an equation for the surface curvature

$$\frac{dh}{dr} = \frac{u_\theta^2}{r \cdot g} \tag{2-28}$$,

based on the pressure profile, which is constant along the surface. Inserting eq. 2-27 in the equation for the surface curvature and solving the differential gives an equation for the fluid level h, which is only depending on the radius r.

From this equation, the theoretically occurring gas core length

$$L = \lim_{r\to\infty} h(r) - h(0) = \frac{\ln(2)}{g}\left(\frac{\Gamma_\infty}{2\pi r_0}\right)^2 \quad (2\text{-}29)$$

can be calculated from the bulk circulation and the vortex core radius [Ito10]. Therefore, with the knowledge of the azimuthal velocity profile, either gained through experiments or numerical simulations, the modelling parameters a and Γ_∞ can be calculated and used to estimate the theoretically occurring gas core lengths.

2.2.4 Granger model

For his dimensionless model, *Granger* used a series expansion of the dimensionless circulation

$$\bar{\Gamma} = \frac{\Gamma}{\Gamma_\infty} \quad (2\text{-}30)$$

and the dimensionless stream function

$$\bar{\Psi} = \frac{\Psi}{Q_r \cdot h} \quad (2\text{-}31),$$

with Q_r as the radial volume flow rate per unit length, to derive an approximation for a stationary, axis symmetrical vortex from the Navier-Stokes equations (eq. 2-11 to 2-13). For the series expansion a dimensionless radial component

$$\eta = \left(\frac{r}{r_0}\right)^2 \quad (2\text{-}32),$$

a dimensionless axial component

$$\xi = \frac{z}{h} \quad (2\text{-}33)$$

and a local radial Reynolds number

$$Re_r = \frac{Q_r}{\nu} \quad (2\text{-}34),$$

are introduced by *Granger*. Assuming that the radial velocity reaches its maximum near the viscous core region and that it is much smaller than the azimuthal and the axial velocity, the series expansion of the circulation

$$\bar{\Gamma}(\eta,\xi) = \bar{\Gamma}_0(\eta,\xi) + \bar{\Gamma}_1(\eta,\xi)Re_r + \bar{\Gamma}_2(\eta,\xi)Re_r^2 + \cdots \quad (2\text{-}35)$$

and the stream function

$$\bar{\psi}(\eta,\xi) = \bar{\psi}_0(\eta,\xi) + \bar{\psi}_1(\eta,\xi)Re_r + \bar{\psi}_2(\eta,\xi)Re_r^2 + \cdots \tag{2-36}$$

can be made. Substituting these into the Navier-Stokes equations and equating the exponents for the radial Reynolds number, results in several equations based on the order of the Reynolds numbers exponent.

The 0. order results in two independent, homogeneous partial differentiable equations

$$2\eta \frac{\partial^2 \bar{\Gamma}}{\partial \eta^2} + \frac{1}{2}\left(\frac{r_0}{h}\right)^2 \frac{\partial^2 \bar{\Gamma}}{\partial \xi^2} = 0 \tag{2-37}$$

$$8\eta^2 \left(\frac{\partial^3 \bar{\psi}}{\partial \eta^3} + \eta \frac{\partial^4 \bar{\psi}}{\partial \eta^4}\right) + \left(\frac{r_0}{h}\right)^2 \left(4\eta^2 \frac{\partial^4 \bar{\psi}}{\partial \eta^2 \partial \xi^2} + \frac{1}{2}\left(\frac{r_0}{h}\right)^2 \eta \frac{\partial^4 \bar{\psi}}{\partial \xi^4}\right) = 0 \tag{2-38}$$.

These equations represent a $2D$ rotational flow, derived from the superpositioning of a solid-body rotation and a potential vortex. The practical applicability of the model is limited by the requirement of small radial Reynolds numbers ($Re_r < 1$) for the ability to solve the series expansion [Gra66].

2.3 Empirical Correlations for the Critical Submergence

Unlike the theoretical vortex models from chapter 2.2, empirical correlations offer a more practical approach to prevent gas entrainment, albeit on the cost of accuracy. A lot of different correlations have been developed within the past decades, of which the most important ones are described in the following. All correlations described in the following, use the Froude number, with varying extensions, to calculate the critical submergence depth S_{crit}.

2.3.1 Jain et al.

The correlation of *Jain et al.* is based on a series of experiments conducted in 1978, investigating the vortex development in cylindrical reservoirs of different sizes and for different fluids. The critical submergence

$$K\frac{S_{\text{crit}}}{D} = 5.6 \cdot \Gamma_{\text{S}}^{*0.42} Fr^{0.5} \tag{2-39}$$

is based on the modified dimensionless circulation

$$\Gamma_{\text{S}}^{*} = \frac{\Gamma \cdot S}{Q} \tag{2-40}$$,

which uses the submergence depth S rather than the intake diameter D (see eq. 2-4) and the Froude number Fr. The factor

$$K = f(\nu^{*}) \tag{2-41}$$,

is a function of the dimensionless viscosity

$$\nu^{*} = \frac{\sqrt{g \cdot D^{3}}}{\nu} \tag{2-42}$$

and only becomes constant for $\nu^{*} \geq 5 \cdot 10^{4}$ [Jai78]. A major drawback of the correlation is the limited practicability due to the many parameters, which need to be known beforehand. Furthermore, numerical simulations conducted by *Pandazis & Blömeling* show that the correlation over predicts the critical submergence for small circulations [Pan13].

2.3.2 Knauss

In his book: *"Wirbelbildung an Einlaufbauwerken – Luft- und Dralleintrag"* from 1983, *Knauss* compares the research of *Anwar et al.* [Anw78], *Anwar & Amphlett* [Anw80], *Dagett & Keulegan* [Dag74] and *Jain et al.* [Jai78]. He unifies them by introducing a dimensionless spin parameter (*German:* Drallparameter)

$$C_{\text{spin}} = \frac{c}{\sqrt{g \cdot D^3}} = \frac{\pi}{4} \Gamma^* \cdot Fr \tag{2-43},$$

which describes the hydraulic and geometric values of the intake chamber, with the spin constant

$$c = u_\theta \cdot r \tag{2-44}.$$

Knauss also introduces a factor for the orientation of the pump intake

$$k = \frac{112.5}{1 + \frac{\varphi}{2\pi}} \; with \begin{cases} \varphi = 0\text{: vertical downwards} \\ \varphi = \frac{\pi}{2}\text{: horizontal} \\ \varphi = \pi\text{: vertical upwards} \end{cases} \tag{2-45}.$$

The resulting equation for the critical submergence hence has the following form

$$\frac{S_{\text{crit}}}{D} = k \cdot C_{\text{spin}} = k \cdot \frac{\pi}{4} \Gamma^* \cdot Fr \tag{2-46}.$$

The correlation is restricted to medium circulations of $C_{\text{spin}} < 0.15$ and Froude numbers between $Fr = 0.33$ to 2 [Kna83]. According to simulations performed by *Pandazis & Blömeling*, the correlation underestimates the critical submergence depth for $Fr < 1.4$ and an intake diameter of $D^{\text{CFD}} = 50$ mm, while it overestimates the submergence depth for higher Froude numbers and intake diameters [Pan13].

2.3.3 ANSI

A correlation, which has been validated for a huge set of geometrical parameters and volume flows, is the correlation given in the *Hydraulic Institutes "American National Standard for Rotodynamic Pumps"* (ANSI)

$$\frac{S_{\text{crit}}}{D_{\text{Intake}}} = 1 + 2.3 \cdot Fr_{\text{Intake}} \tag{2-47},$$

which is based on the correlation developed by *Hecker* [Hec78] but does not include any circulation. In the ANSI correlation, S_{crit} is only depending on the intake opening diameter D_{Intake} and the intake Froude number Fr_{Intake}, also using D_{Intake} instead of the intake pipe diameter D. Since the correlation does not include the circulation, the critical submergence can be derived, without knowledge of the flow conditions inside the intake chamber. The applicability of the model is limited to flows with "low to moderate circulations", using the intake velocity and the velocity of the approaching flow as limitations, rather than the circulation itself. The values are $0.6\ \text{ms}^{-1} < u_{\text{intake}} < 2.7\ \text{ms}^{-1}$ for intake and $0.9\ \text{ms}^{-1} < u_{\infty} < 1.2\ \text{ms}^{-1}$ for approaching flow velocities, with exact values depending on the geometry and size of pump intake and intake chamber. If the intake velocity is above the limit for the specific intake geometry, the installation of vortex breakers into the pump intake is recommended by the *Hydraulic Institute*. Pump systems with intake volume flow rates $Q > 315\ \text{Ls}^{-1}$ require verification studies with a physical model [Hyd12]. According to *Pandazis & Blömeling*, the ANSI correlation underestimates the critical submergence by a large margin. Only for small diameters ($D^{\text{CFD}} = 20.75$ mm) and high Froude number ($Fr > 15$) the correlation is in agreement with the numerical results [Pan13].

The correlation can also be rearranged, to derive a critical Froude number

$$Fr_{\text{crit}} = Fr(S_{\text{crit}}) = Fr(L^* = 1) \qquad (2\text{-}48),$$

the Froude number at which gas-core length is equal to the submergence depth and the tip of the gas-core reaches the pump intake.

2.3.4 Additional correlations

Next to the three correlations described in the previous subchapters, there exist many other correlations, i.e. the ones of *Rindels & Gulliver* [Gul83] and *Odgaard* [Odg86]. The equations and limitations for the correlations can be found together with the ones of *Jain et al.*, *Knauss* and ANSI in Table 2-1. There exist also correlations for rotation-free vortices (see chapter 2.1), namely the correlations of *KSB* [Wag05] and *Lubin & Springer* [Lub67], but as the investigation of rotation-free vortices is not a part of this thesis, they are not further regarded.

Table 2-1: Correlations for the critical submergence.

Correlation	Equation	Intake orientation	Application range
ANSI [Hyd12]	$\frac{S_{\text{crit}}}{D_{\text{Intake}}} = 1 + 2.3 \cdot Fr_{\text{Intake}}$	var.	$0.6\ \text{ms}^{-1} < u_{\text{intake}} < 2.7\ \text{ms}^{-1}$ $0.9\ \text{ms}^{-1} < u_{\infty} < 1.2\ \text{ms}^{-1}$ $\Gamma \sim$ moderate
***Jain et al.* [Jai78]**	$K \frac{S_{\text{crit}}}{D} = 5.6 \cdot \Gamma_S^{*0.42} Fr^{0.5}$	↓	$\nu^* \geq 5 \cdot 10^4$ $1.1 < Fr < 20$
***Knauss* [Kna83]**	$\frac{S_{\text{crit}}}{D} = k \cdot \frac{\pi}{4} \Gamma^* \cdot Fr$	var.	$C_{\text{Spin}} < 0.15$ $0.33 < Fr < 2$ $S_{\min} > 1.5 \cdot D$
***Odgaard* [Odg86]**	$\frac{S_{\text{crit}}}{D} = 7.5 \cdot Fr\sqrt{\Gamma^*}$	↓	$We > 720 \cdot \sqrt{\frac{Fr}{\Gamma^{*3} \cdot \sqrt{Re}}}$
***Rindels & Gulliver* [Gul83]**	$\frac{S_{\text{crit}}}{D} = 2.5 + \frac{4}{3} Fr^{\frac{2}{3}} + 40\alpha^{*3}$ $\alpha^* = \frac{\tan\alpha}{1 + (L/B)\tan\alpha}$	↓	$0.25 < Fr < 3$

2.4 Scaling Criteria

Scaling describes the process of changing the size of a system. It can either be done as a scale-up from an experimental plant to a full-size industrial plant or as a scale-down from an industrial size plant to a model plant for verification experiments. Scaling needs to be done according to a set of scaling rules, otherwise the resulting systems might not be comparable to the original.

The first scaling criteria is the geometric similarity. Ideally, the smaller system is just a linearly shrunken model of the larger system, with all lengths scaled down by the same factor. The second scaling criteria is the dynamic similarity. The purpose of this criteria is, to keep all internal and external forces of the system within the same range. Especially fluid dynamic parameters, like velocity, acceleration, viscosity and surface tension, have a great impact on the dynamic similarity. The dynamic similarity is much harder to keep, as the relevant parameters are scaling by different factors or even not at all and certain thresholds must be considered, i.e. laminar or turbulent flow conditions. When geometric and dynamic similarity criteria are met, the third scaling criteria, the kinematic similarity, is also met.

Table 2-2: Primary quantities and their dimensions [Zlo06].

Physical Parameter	Dimension	Unit
Length	L	m
Mass	M	kg
Time	T	s
Temperature	Θ	K
Molecular amount	N	mol
Electric current	I	A
Luminous intensity	J	cd

For a successful scale-up a dimensional analysis can be used. In a first step all variables influencing the system are gathered and have their dimension recorded. The dimension describes the physical properties of a variable, i.e. Length (L), Mass (M) or Time (t). A dimension is purely qualitative and carries no quantitative value, i.e. whether an object is long or short, heavy or light etc. All variables can be separated into two categories: Primary (or base) and secondary quantities.

Zlokarnik defines the primary quantities as those quantities based on the seven SI units. The secondary quantities include all other quantities, which are derived from the primary quantities, i.e. volume, velocity or pressure. In Table 2-2 all primary and in Table 2-3 the important secondary quantities for this thesis are written down.

After all influencing parameters are recorded, a target quantity must be defined, and the system must be made dimensional homogeneous. Dimensional homogenous means that the target quantity is dimensionless and all parameters which influence the target quantity are linear independent of each other. Often, a known dimensionless number can be selected as a target quantity, like the Reynolds, Weber or Froude number.

Table 2-3: Secondary quantities and their dimensions.

Physical Parameter	Dimension	Unit
Velocity	$L \cdot T^{-1}$	ms^{-1}
Acceleration	$L \cdot T^{-2}$	ms^{-2}
Density	$M \cdot L^{-3}$	kgm^{-3}
Volume	L^3	m^3
Area	L^2	m^2
Momentum	$M \cdot L \cdot T^{-1}$	$kgms^{-1}$
Pressure	$M \cdot L^{-1} \cdot T^{-2}$	$Pa = kgm^{-1}s^{-2}$
Circulation	$L^2 \cdot T^{-1}$	m^2s^{-1}
Surface tension	$M \cdot T^{-2}$	kgs^{-2}
kinematic viscosity	$L^2 \cdot T^{-1}$	m^2s^{-1}

2.4.1 Froude Scaling

The Froude scaling is a scaling method based on geometrical similarity, where the Froude number is the selected target quantity. The method is used by *Knauss* and the *Hydraulic Institute* [Kna83], [Hyd12]. In keeping the Froude number constant between the different scales

$$Fr = \frac{u_1}{\sqrt{g \cdot D_1}} = \frac{u_2}{\sqrt{g \cdot D_2}}, \tag{2-49}$$

the velocities can be expressed by

$$u_2 = u_1 \sqrt{\frac{D_2}{D_1}} = u_1 \sqrt{\lambda}, \tag{2-50}$$

which gives a geometrical scaling factor

$$\lambda = \frac{D_2}{D_1} \tag{2-51}$$.

2.4.2 Scaling limitations for vortex investigations

For the scaling of vortex systems, the relevant dimensionless numbers are listed in chapter 2.1.1. As it is impossible to keep all numbers constant, several authors have found limiting values for scaling. The Froude number, the dimensionless circulation and the submergence depth to intake diameter ratio (S/D) are the critical scaling parameters for the vortex investigations and should be kept constant over all scales [Kna83]. The effect of the viscosity on the system is described by the Reynolds number. The investigations show that the influence of the viscosity decreases for large Reynolds numbers and therefore can be neglected above a certain limit.

Jain et al. and *Daggett & Keulegan* have found the following threshold for systems with vertical downwards pumping intakes:

Jain et al.	1978	$Re > 3.2 \cdot 10^4$
Daggett & Keulegan	1978	$Re > 5 \cdot 10^4 \cdot Fr$

For horizontal or vertical upwards facing pump intakes *Anwar & Amphlett* found that

Anwar & Amphlett	1980	$Re > 3.8 \cdot 10^4 \cdot \frac{S}{D}.$

The *Hydraulic Institute* considers a threshold of

ANSI	2012	$Re > 6 \cdot 10^4,$

to be sufficient to neglect the influence of the viscosity, regardless of pump intake orientation. While Knauss recommends an even higher value of

Knauss	1983	$Re > 10^5.$

All these values are recommendations, as there is no clear border for the influence of viscosity and should be treated as such.

For the influence of the surface tension the Weber number can be used. *Jain et al., Odgaard* and the *Hydraulic Institute* recommend as thresholds for the Weber number

Jain et al.	1978	$We^2 > 120$
Odgaard	1986	$We > 720 \cdot Fr^{0.5} \cdot \Gamma^{*-\frac{3}{2}} \cdot Re^{-\frac{1}{4}}$
ANSI	2012	$We^2 > 240$

The recommendation values of the Hydraulic Institute for the Reynolds and the Weber number are based on *Jain et al.* with a security factor of 2. For closed cooling-circuits the Hydraulic Institute recommends a minimal value for the Reynolds number of $Re > 10^5$ [Hyd12].

For geometrical scaling, *Anwar* found that the scaling factor shouldn't exceed

Anwar & Amphlett	1980	$\lambda < 20,$

which is supported by *Dhillon* [Dhi80]. Further, *Knauss* recommends a minimal intake diameter of

Knauss	1983	$D \geq 75$ mm

for comparable results with larger scale systems, based on the work of *McCorquodale* [Kna83].

2.5 Vortex Prevention

Since it is not always possible to maintain a secure submergence depth, due to constructive reasons or temporary water shortages, vortex breakers can be used to hinder the formation of vortices and to prevent gas-entrainment. Vortex breakers are static installations which can be implemented into or above the pump intake. For the experiments conducted in this thesis, the vortex breakers are divided into three different categories, based on their working principles. The first category consists of vortex breakers designed to reduce the intake velocity and therefore the vortex strength, by changing the size and shape of the pump intake opening. Vortex breakers of this category include bell-mouths and funnels. Important design parameters are size, opening angle and shape of the edge [Yan14]. The second category includes vortex breakers, which reduce the circulation above the pump intake through baffles and sieves. Several authors propose the implementation of static vortex breakers, see *Auckland* [Auc09], *Trivellato* [Tri10], *Borgei & Kabiri-Samani* [Bor10] as well as the *Hydraulic Institutes "American National Standard for Rotodynamic Pumps"* [Hyd12]. Without circulation, the development of a stable vortex core is hindered and the pressure loss, responsible for the gas core formation is reduced. Vortex breakers of this category are cross-shaped structures or perforated baskets. The third category are baffle shaped vortex breakers, installed above the pump intake to block a direct path from the liquid surface to the pump intake. Baffles should not be installed too close to the intake, the *Hydraulic Institute* recommends a distance of $D/2$, as otherwise the absolute pump intake area decreases. The optimal diameter of the baffle plate is $4D$, according to the *Hydraulic Institutes* guidelines [Hyd12].

It is possible to combine two types of vortex breakers, to improve the overall performance. Common types are combinations of crosses and baffles, see Figure 2-9 for visualization. Based on the work of *Padamanabhan* [Pad82], the *Hydraulic Institute* recommends horizontal baffles, crosses or cross-baffle combinations implemented above the pump intake to suppress the vortex formation [Hyd12].

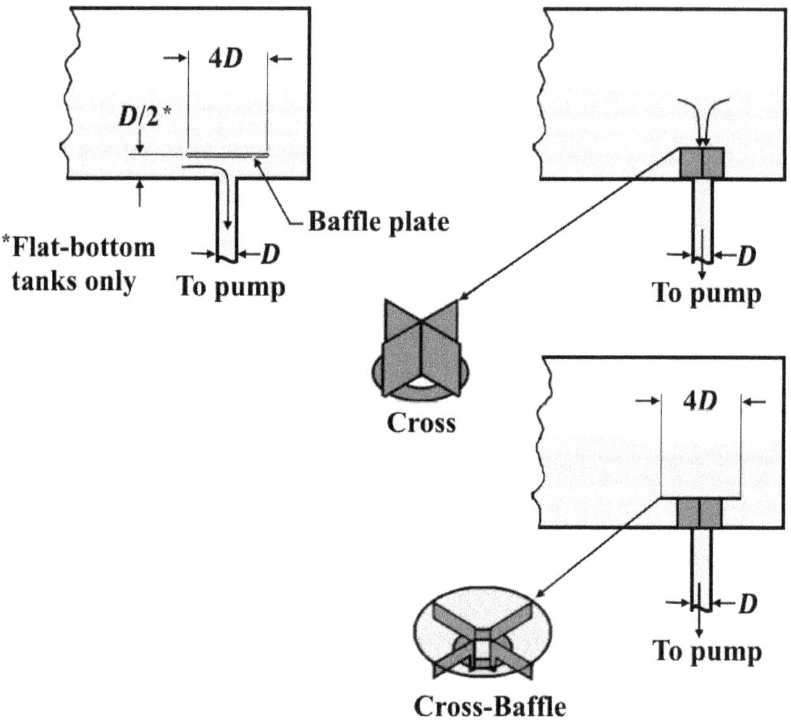

Figure 2-9: Vortex breakers in a vertical pump intake. Clockwise: Baffle plate, cross and baffle-cross, drawing according to [Hyd12].

In chemical industries, vortex breakers are recommended to be used in storage tank outlets, to prevent a pressure loss and gas-entrainment into the system [Kis90, Lud99, Mah10 & Man95], as well as cyclones, to increase separation efficiency [Oga00]. Although general rules of design (the proverbial rule of thumb, see Figure 2-10) are given for tank outlet vortex breaker, no evaluation regarding the efficiency of the different vortex breaker designs are published to this author's knowledge. This is insofar notable, as several patents for vortex breaker have been approved in the past decades, like [Ekh12, Lis00, Mor09, Rah87 & Stri92].

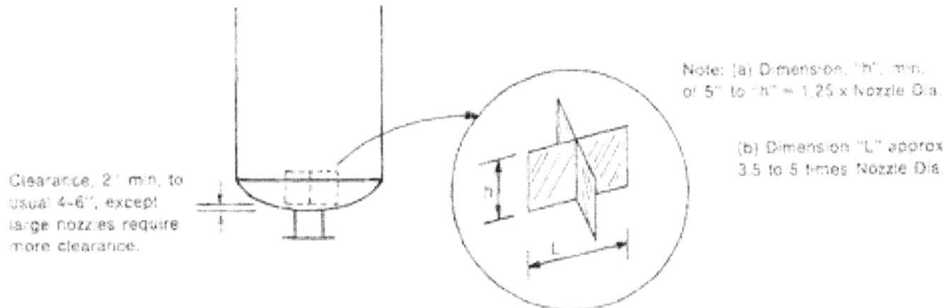

Figure 2-10: Design recommendation for vortex breakersin tank outlets [Lud99].

3 Experimental Setup

In this chapter the overall conditions, different setups used, and experiments conducted in the globe of this thesis, are described in detail. Notable measuring apparatuses used in the experiments are additionally described shortly in the end of the chapter.

3.1 General Proceedings

3.1.1 Experimental Conditions

The water used in the experiments is de-ionized (DI) water for the DN15 laboratory plant and tab water for the DN200 pilot plant. The water temperature in the DN15 laboratory plant is kept constant at 20 ± 2 °C by cooling with a closed cooling circuit, using the universities cooling water supply, and heated up when necessary, by using the dissipation energy of the pump. Although the DN200 pilot plant is also connected to the cooling water system through a tubular heat exchanger, normally no cooling is necessary, since the heat capacity of the water and the heat loss across the steel surface of the setup lead to a constant temperature during the experiments. Nonetheless, the temperature is monitored during the experiments, to prevent measurement uncertainties. The experimental conditions and general properties are listed in Table 3-1.

The experimental setups are named after their respective sizes, laboratory and pilot plant, with the number in the beginning, DN15 and DN200, indicating the nominal diameter of the pump intake and peripheral piping. The setups are designed to be geometrically similar, with a scaling factor of 13.3 between the pump intake diameters and submergence depths. Both are operated within the same range of Froude numbers. Detailed descriptions of each setup are given in the following chapters 3.1.2 and 3.1.3.

Table 3-1: Experimental conditions and properties.

Parameter:	Value:
Operating water temperature T/°C	20 ± 2
Operating pressure p/Pa	$\sim 10^5$
Kinematic viscosity† (H_2O, 20 °C) ν/m²s⁻¹	$1.004 \cdot 10^{-6}$
Density† (H_2O, 20 °C) ρ/kgm⁻³	998.21
Surface tension† (H_2O-Air, 20 °C) σ/Nm⁻¹	$72.75 \cdot 10^{-3}$

†: Fluid properties are taken from the VDI Heat Atlas [VDI10]

3.1.2 DN200 Pilot Plant

The pilot plant setup is located inside the research facilities of the Institute of Multiphase Flows at Hamburg University of Technology (TUHH). The plant consists of a cylindrical pump reservoir made from coated carbon steel, while the piping consists of PVC-U pipes with a diameter of DN200 (PN10). The reservoir has a diameter D_R and height H_R of 4 m each and can hold a total volume of $V = 50$ m³. Optical accessibility into the reservoir is given by 24 round borosilicate glass windows, which are equally distributed in three rows around the circumference of the reservoir. For photographs of the setup, see Figure 3-1.

Figure 3-1: DN200 pilot plant setup inside the experimental hall of the TUHH. a) ABB FEX311 electromagnetic flowmeter, b) Pump – KSB MegaCPK, c) Pump reservoir.

The pilot plant is operated as a loop flow, with a vertical downward pump intake installed inside the reservoir at a height of $h = 1.25$ m. From the intake, the water is flowing through a 90°-bend and a horizontal pipe to the installed centrifugal pump KSB-MegaCPK, see Figure 3-1b (for the data sheet, see the Appendix). From there the water is pumped upwards, through a tubular heat exchanger and split evenly into four separate streams. Each of these streams is measured by an electromagnetic flowmeter of the type ABB FEX311, before each stream re-enters the reservoir. The four entry pipes are arranged in a circular array around the circumference of the reservoir, see Figure 3-2 for a 3D-flowchart and a cross-sectional view of the setup and Table 3-2 for the dimensions.

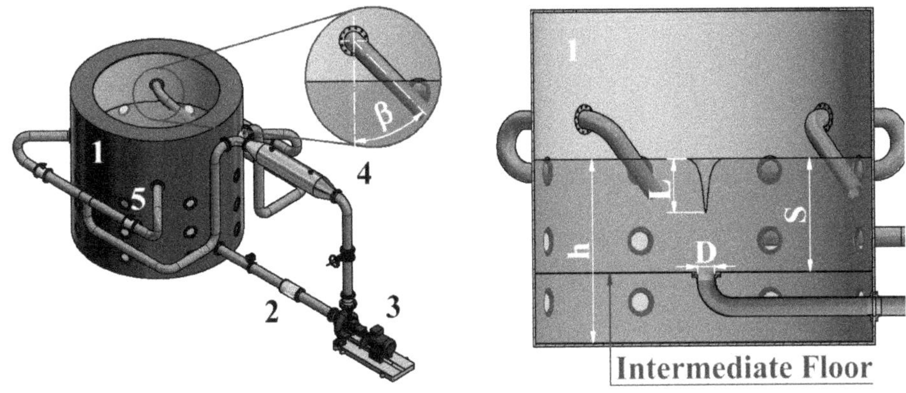

Figure 3-2: 3D-flowchart of the DN200 pilot plant setup. Left: Isometric view, right: Cross-sectional view of the cylindrical reservoir. 1: Pump reservoir, 2: Optical access point – suction pipe, 3: Pump – KSB MegaCPK, 4: Shell and tube heat exchanger, 5: Electromagnetic flow meter - ABB FEX311 (4x).

Table 3-2: Dimensions of the DN200 pilot plant setup reservoir.

Parameter:	Value:
Reservoir height /m	4.0
Reservoir diameter D_R/m	4.0
Pump intake diameter D_{Intake}/m	0.2
Submergence depth above intake S/m	1.07 – 2.13
Vertical inflow angle β/°	15 & 45
Height of intermediate floor h_{floor}/m	1.25

Inside the reservoir the inflow pipes are bent to face the wall in a defined, vertical angle β. For most experiments, the inflow angle is $\beta = 45°$. For one set of experiments, the inflow angle was changed to $\beta = 15°$. The pipe ends are submerged, to avoid bubble generation in the inflow. On the highest point of each inflow pipe, a ¼" hose is installed, which connects to a vacuum pump. This setup is used to remove gas-pockets from the pipe and therefore guarantees that each pipe has the same effective volume flow rate. For selected experiments an intermediate floor is installed over the whole cross-section of the tank at the same height as the pump intake. The intermediate floor creates a higher degree of geometrical similarity to the laboratory sized experiments and makes the system above the floor more symmetrical, which is especially important for the optical measurements of the velocity fields.

3.1.2.1 Intake Shapes and Vortex Breakers

The pump intake is an open tube, with sharp edges, enabling the installation and investigation of different vortex breakers or intake shape variations. The installed vortex breakers and intake shapes can be seen in Figure 3-3. Six different vortex breakers are investigated in the DN200 setup: Cross, Flat Cross, Flat Baffle-Plate, Cross-Plate and Large Cross-Plate. Additionally, four intake shape variations are also installed and investigated: Open Tube, Funnel, Sieve, and 45°-Bend.

The funnel shape is folded from stainless steel and welded to a flange, also made from stainless steel. The opening angle of the funnel is 45°, the smaller diameter is the same size as the pump intake ($D_{Intake} = 0.2$ m), while the top diameter is $d = 0.35$ m, which results in an opening area of 0.1 m^2, which is three times as much as the open tube intake. The sieve structure consists of galvanized steel, in which 28 holes with a diameter of 26.6 mm each are punched. The holes are arranged circular on three different radii. The total opening area of the sieve is 0.016 m^2, which is about one-half of the open tube intake opening area. The 45°-bend is made from a PVC-U pipe and has the same opening diameter and area as the open tube intake. The intake areas and effective diameters for the intake shapes are listed in Table 3-3.

All vortex breakers have a height of $h = 0.1$ m above the pump intake, equaling in $h = 0.5 \cdot D_{Intake}$. Two different designs can be distinguished: One set of vortex breakers is made from transparent acrylic-glass, while the second set of vortex breakers is made from

PVC-U. The acrylic-glass set consists of two different structures, a cross and a baffle-plate vortex breaker. The cross has a total height of h = 0.3 m, equaling $h = 1.5 \cdot D_{Intake}$, with two-thirds of the height being inside the pump intake, where the cross is as wide as the pipe. Above the intake, the cross has a diameter of d = 0.4 m or twice the pump intake diameter ($d = 2D_{Intake}$). The baffle-plate consists of four supporting baffles, arranged in a cross-shape, on which the plate is mounted. The plate also has a diameter of twice the pump intake ($d = 2D_{Intake}$), while the supporting baffles have a height of h = 0.1 m. For both structures the acrylic glass used has a thickness of b = 5 mm.

For the PVC-U vortex-breakers, the basic structure is a cross shaped vortex breaker, with the same dimensions as the one made from acrylic-glass. The material also has a thickness of b= 5 mm. The cross-plate structures are constructed by mounting plates of different sizes ($d = 2D_{Intake}$ and $d = 4D_{Intake}$) onto the cross-structure.

The different intake shapes and the acrylic-glass vortex breakers are investigated with the intermediate floor is installed, while the PVC-U vortex breakers are investigated without the use of the intermediate floor. All structures are implemented by fixing them with bolts onto the pump intake.

Table 3-3: Intake areas and diameter of the investigated intake shapes for the DN15 laboratory and the DN200 pilot plant.

Setup	**Intake Shape**	**Intake Area A_{Intake}/m^2**	**Effective Intake Diameter D_{Intake}/m**
DN15	Open Tube	$1.77 \cdot 10^{-4}$	0.015
	Funnel	$7.07 \cdot 10^{-4}$	0.030
	Sieve	$0.88 \cdot 10^{-4}$	0.011
	Recess	$0.88 \cdot 10^{-4}$	0.011
DN200	Open Tube	0.031	0.2
	Funnel	0.10	0.355
	Sieve	0.016	0.1
	45°-Bend	0.031	0.2

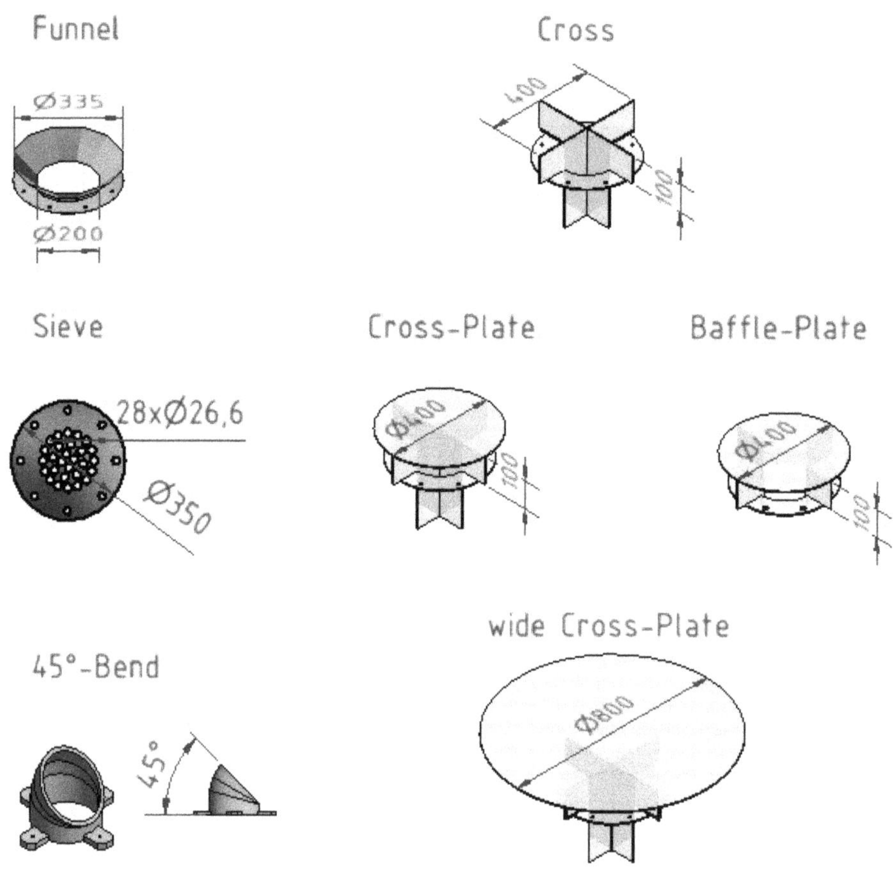

Figure 3-3: Intake shapes and vortex breakers in the DN200 pilot plant setup, with the most important sizes. Not shown: Open Tube intake shape, which is the default shape.

3.1.3 DN15 Laboratory Plant

The laboratory plant setup is erected on a mobile worktop within the laboratories of the *Institute of Multiphase Flows* at *Hamburg University of Technology* (TUHH). Like the DN200 pilot plant setup, its main component is a cylindrical vessel, functioning as a pump intake, with vertical downwards facing outflow. The vessel body is made from acrylic glass pipe with a diameter of DN300 ($D_R = 0.288$ m), see Figure 3-4a for photographs of the laboratory setup. The vessel bottom as well as the peripheral piping and installations are changed during the course of this thesis, but the geometric similarity to the pilot plant setup is kept constant. Therefore, the setup is always operated as a loop flow with pump intake and piping diameter of DN15 and four entry pipes circularly arranged around the circumference of the reservoir for the backflow. See Figure 3-4b for a flowchart of the setup.

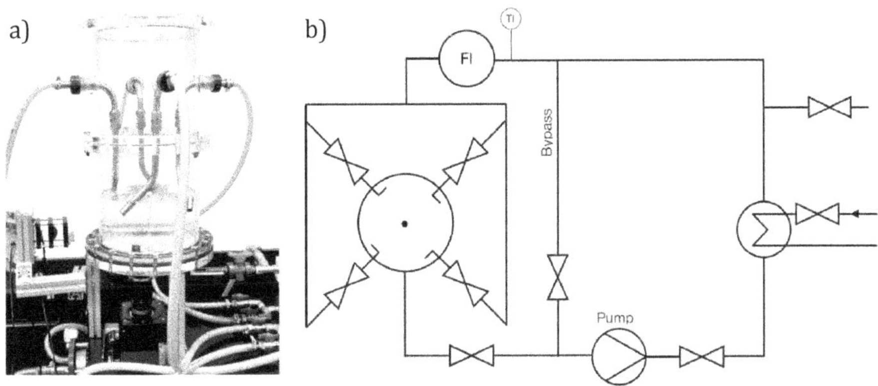

Figure 3-4: Picture and flowchart of the DN15 laboratory setup. a) Picture of the cylindrical vessel with transparent bottom and four inflow pipes ($\beta = 45°$), high-power LED visible on the left side. b) Flowchart of the setup, showing pump, heat exchanger, bypass, flow meter and valves.

A circular pump is used, and the volume flow and temperature are measured with an E+H Promass80F flow meter with included temperature probe. The heat exchanger for the temperature control is located directly behind the flow meter. The vessel bottom, including the pump intake is made from PVC-U during the first experiments and later exchanged against a transparent acrylic glass bottom for optical accessibility. The inflow pipes are made from stainless steel. Different sets of pipes exist with predefined vertical angles of β = 45°, 30°, 15° and 0°, where 0° equals a vertical downwards facing opening. Each inflow pipe can be manually opened and closed with a ball valve, which is used to remove accumulated gas from each pipe.

In the experiments for the determination of the vortex flow field, a transparent rectangular box is installed around the vessel and filled with DI-water, to reduce the distortions caused by the vessels curved surface and for refractive index matching.

3.1.3.1 Intake Shapes and Vortex Breakers

The intake shapes are cut from PVC-U and consist of a flat disk (d = 59 mm, h = 8 mm), with a thread on the outside and the intake in the center, see Figure 3-5. The disks can be screwed into the PVC-U bottom of the DN15 laboratory plant, which has a fitting inner thread in the middle. For the vortex breakers, the acrylic-glass bottom is used, which has a fixed, cylindrical pump intake. The vortex breakers are installed into the pump intake by pressing them directly into the tube of the intake. All breakers are designed with CAD and manufactured with rapid prototyping from acrylic resign. See Figure 3-6 for CAD models of the vortex breakers.

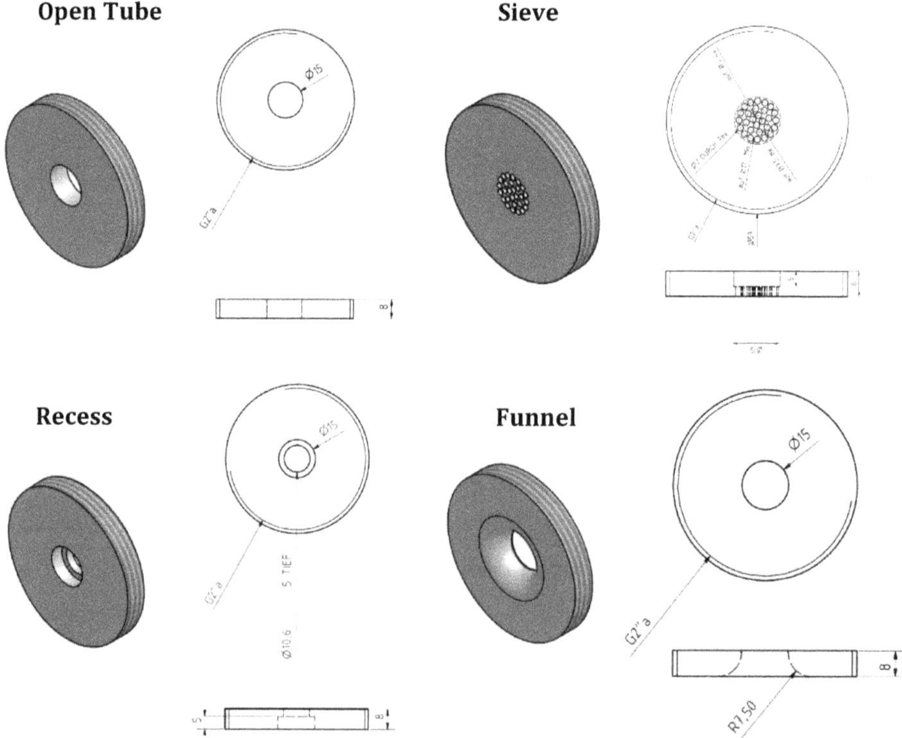

Figure 3-5: Investigated intake shapes in the DN15 laboratory plant.

Ten different vortex breakers are investigated in the DN15 setup, of which three are geometrical similar to the ones in the DN200 setup (marked with a '*'): Inner-Cross, Cross*, high Cross, high and wide Cross, Baffle-Plate*, wide Baffle-Plate (two different variations), Cross-Plate*, high Cross-Plate and Bell. Additionally, four intake shape variations are also installed and investigated: Open Tube, Funnel, Sieve, and Recess. While the Sieve and the Open Tube are geometrical similar to the intake shapes investigated in DN200 pilot plant, the funnel is not protruding into the water and has an opening area of four times the open tube intake instead of three as in the DN200 pilot plant. Instead of a 45°-bend a recess shape is investigated, which narrows the pump opening to d = 10.6 mm and therefore creates the same effective opening area as the sieve.

The vortex breakers are of the same basic shapes as described in chapter 3.1.2, except the bell and inner-cross shape, which can be seen in Figure 3-6. The bell-shape is designed to divert the inflow to the intake several times. Therefore, the bell-shaped vortex breaker has roughly the same shape as the cross-plate structure, but along the circumference of the plate it has a rim, which reaches down 3/4th of the height. Inside the structure, the cross is smaller ($d = D_{Intake}$) and around the pump intake is another rim, which reaches from the bottom to the middle of the structure.

The inner-cross structure on the other hand consists of the lower part of the cross structure, without protruding parts from the pump intake.
Of the wide baffle-plate, two variations are investigated, to distinguish the influence of the baffles on the vortex suppression. The default type has short baffles, with a diameter of $2D_{Intake}$, while the variation of the structure has wide baffles with a diameter as wide as the plate on top ($4D_{Intake}$).

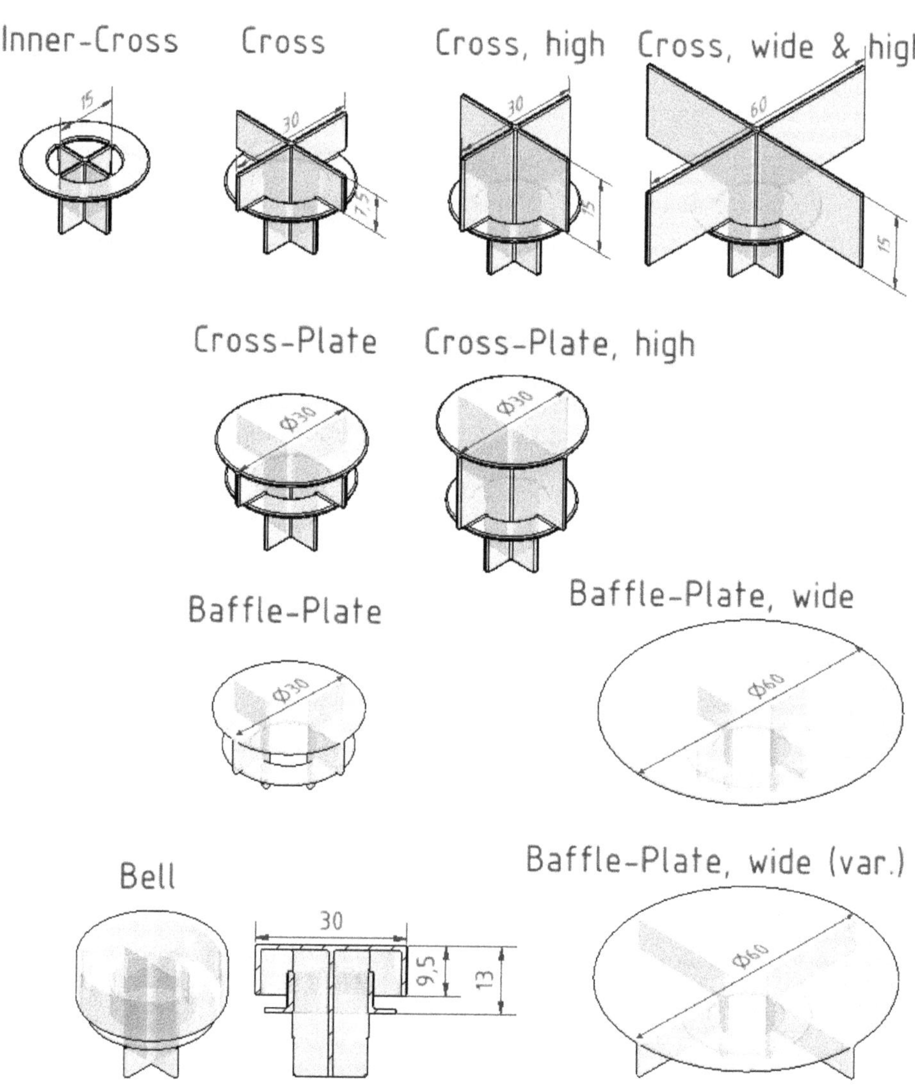

Figure 3-6: Vortex breakers investigated in the DN15 laboratory plant.

3.2 Conducted Experiments

The experiments conducted in this thesis can be separated into two main parts: Investigation of the Gas-Core Lengths and Particle Image Velocimetry Measurements. Both parts can be divided further by the two different setup sizes and the conducted experiments. Chapter 3.2.1 consists of the experiments conducted both in the DN200 pilot plant and the DN15 laboratory plant to determine the effects of different operative and constructive measures on the development and strength of a gas-core above the pump intake. Measures include variation of submergence depth, volume flow rate, induced momentum, intake shape and size as well as the installation of vortex breakers. For a detailed list of all volume flow rates and the corresponding velocities and dimensionless numbers, see table A-1 and A-2 in the Appendix.

In chapter 3.2.2 and 3.2.3 the flow conditions above and inside the pump intake are investigated, using dye-core experiments and (laser-)optical measurement techniques, namely Particle Image Velocimetry.

3.2.1 Investigation of the Gas-Core Lengths

The gas-core lengths are measured, using a shadowgraph imaging technique. Therefore, a DLSR (digital single-lens reflex) camera is mounted in front of one of the access windows in case of the DN200 pilot plant, and in front of the transparent vessel in case of the DN15 laboratory plant. For the DN200 pilot plant, a 1000 W spotlight is orthogonally aligned to the camera in front of another access window, while for the DN15 laboratory plant an LED panel is used as a backlight.

Experiments are conducted first for an open tube pump intake without vortex breakers or intake shape variations at three different submergence depths $S^{\mathrm{PP}} = 1.1, 1.5, 2.1$ m and Froude numbers ranging from $Fr^{\mathrm{PP}} = 0.4$ to 1.8 for the DN200 pilot plant as well as $S^{\mathrm{LP}} = 0.08, 0.11, 0.16$ m and Froude numbers ranging from $Fr^{\mathrm{LP}} = 0.2$ to 1.8 for the DN15 laboratory plant. The dimensionless submergence depths

$$S^* = \frac{S}{D} \tag{3-1}$$

are similar for both setups, with values of $S^* = 5.3, 7.3$ and 10.7. For all experiments in the DN200 pilot plant, the angle of inflow is $\beta = 45°$, except for one measurement row with $S = 1.46$ m, where the gas-core lengths are also investigated for a vertical inflow angle of $\beta = 15°$. In the DN15 laboratory setup all experiments are conducted for $\beta = 45°$, as well as additional measurements with inflow angles of $\beta = 0°, 15°$ and $30°$. The measurements with vortex breakers are conducted for the same submergence depths and Froude numbers of $Fr^{\mathrm{PP}} = 0.5$ to 4.6 in the DN200 pilot plant and $Fr^{\mathrm{LP}} = 0.5$ to 6.5 in the DN15 laboratory plant. The angle of the inflow pipes is kept for all experiments with vortex breakers at $\beta = 45°$. Table 3-4 lists all experimental parameters for the gas-core measurements in both setups. As the experiments in the DN15 laboratory setup are less time consuming, more vortex breakers are investigated in the small scale.

Before the experiments are started, a target picture is recorded prior to each measurement, to ensure precise results. For the target picture, a metering rod is placed from the surface onto the pump intake, thus enabling a precise measurement of the submergence and of the vertical distortion caused by the camera opening angle and lens. For a further reduction of the measurement error, there is a 15 min waiting period between the formation of the gas-core and the start of the recording in the DN200 pilot plant and a 5 min waiting period in the DN15 laboratory plant, to ensure a stable gas-core has formed and steady-state conditions are reached. All recordings are evaluated with *Matlab*, to automate the process. Individual measurement rows are additionally evaluated manually, using the open-access software ImageJ, to exclude the possibility of errors in the automatic evaluation process. Due to the strong induced momentum, the formed vortices are generally very stable and a sample size of 60 recordings with a recording frequency of $f = 1$ Hz is deemed sufficient to obtain the correct gas-core lengths for each measurement with $\beta = 45°$, for measurements with a smaller inflow angle, 200 recordings are used as a sample size. Each measurement is conducted at least twice and then averaged, to reduce measurement uncertainties and get an error estimation.

Table 3-4: Experimental parameters for the DN200 pilot plant and the DN15 laboratory plant setup.

Parameter:	DN200 Pilot Plant	DN15 Laboratory Plant
Intake volume flow rate $Q/\mathrm{m^3h^{-1}}$	63.3 – 792	0.05 – 1.58
Intake velocity $u/\mathrm{ms^{-1}}$	0.56 – 6.86	0.08 – 2.49
Intake Froude number $Fr/-$	0.4 – 4.6	0.2 – 6.5
Submergence depth above intake S/m	1.07 – 2.13	0.08 – 0.16
Vertical inflow angle $\beta/°$	15 & 45	0, 15, 30, 45
Pump intake diameter $D_{\mathrm{Intake}}/\mathrm{m}$	0.2	0.015
Intake Shapes	Open Tube Funnel Sieve 45°-Bend	Open Tube Funnel Sieve Recess
Vortex Breakers	Cross Baffle-Plate Cross-Plate Cross-Plate, wide	Inner-Cross† Cross Cross, high Cross, high & wide Baffle-Plate Baffle-Plate, wide†† Cross-Plate Cross-Plate, high Bell

Height above intake: normal†††: $h = 0.5 \cdot D_{\mathrm{Intake}}$, high: $h = D_{\mathrm{Intake}}$

Diameter: Normal†††: $d = 2 \cdot D_{\mathrm{Intake}}$, wide: $d = 4 \cdot D_{\mathrm{Intake}}$

†: Inner-Cross has a height above intake of 0 and a diameter of $d = D_{\mathrm{Intake}}$

††: Two different variation of the wide Baffle-Plate are investigated, see Figure 3-6

†††: normal is the default size, which is not extra labeled.

3.2.2 Dye Experiments

For selected experiments, dye core experiments are conducted by inserting blue ink into the vortex center. In the DN200 pilot plant *1 liter* of ink is used, while in the DN15 laboratory plant a few drops of ink are sufficient for the coloring. The coloring enables a qualitative description of the flow structures above the pump intake and a rough estimation of the vortex core radius.

Dye experiments are conducted in both setups for the medium, dimensionless submergence depth of $S^* = 7.3$ ($S^{PP} = 1.46$ m, $S^{LP} = 0.11$ m) and $\beta = 45°$. In the DN200 pilot plant, the experiments are done for the open tube and the cross-baffle *2D* vortex breaker. In the DN15 laboratory plant, the experiments are done for the open tube and five different vortex breakers (Cross-*1D*, Cross-*2D*, Cross-Baffle-*2D*, Cross-Baffle-*4D* & Bell). Additionally, one set of experiments is done for the open tube and an inflow angle of $\beta = 15°$. Investigated Froude numbers vary between the experiments, since for every measurement point the minimal Froude number Fr_{min} and the half-length Froude number $Fr_{0.5}$ are taken. Fr_{min} is the Froude number at which a gas-core starts to form, while $Fr_{0.5}$ is the Froude number at which the gas-core length has reached half the submergence depth ($S^* = 0.5$).

3.2.3 Particle Image Velocimetry Measurements

Knowledge of the flow field within vortices is necessary to gain an understanding in the underlying physical phenomena, which lead to the formation of the vortices. The azimuthal velocity profile and the circulation are thereby the most important values according to the theoretical models, see chapter 2.2. The circulation is also a parameter which is used in many empirical correlations (see chapter 2.3). Therefore, the following experiments are conducted in both setups, to get a qualitative and quantitative description of the flow conditions above the pump intake.

3.2.3.1 Measurement Principle

Particle Image Velocimetry (PIV) is a measurement technique which enables the visualization of velocity fields inside a fluid. To visualize the flow, tracer particles are added to the fluid and illuminated with help of a light source. Often, a laser is used, since the light emitted by a laser can be spread into a thin, almost two-dimensional plane. The movement of the tracer particles in the flow can then be recorded and evaluated for two consecutive images, by converting the recordings into grayscale values and applying a correlation analysis. With knowledge of the conversion rate between pixels and physical lengths and the time between the two recordings, the velocities in x- and y direction can be calculated. While standard PIV uses double image recording for a discrete measurement of the flow field, high-speed PIV can be used to get a continuous recording of the flow. [Raf07]

3.2.3.2 DN200 Pilot Plant Setup

Two sets of PIV measurements are conducted in the DN200 pilot plant for a submergence depth of $S^{\mathrm{PP}} = 1.5$ m ($S^* = 7.3$) and a vertical inflow angle of $\beta = 45°$. The first measurement set is carried out in the bulk region of the vessel at a radius between $r = 0.15 - 0.22$ m, and a measurement height of $h = 0.59$ m above the pump intake. The measurements are conducted for four different Froude numbers $Fr^{\mathrm{PP}} = 0.5, 0.7, 1.0$ and 1.4. A second measurement set is done with a field of view inside the vortex core region at a radius between $r = \pm 0.06$ m, a measurement height of $h = 1.12$ m above the pump intake, and a Froude number of $Fr^{\mathrm{PP}} = 0.5$. The overall setup for both PIV measurements sets is identical and will be explained in detail in the following paragraphs.

As a light source a newly developed High-Power LED is used, while polyamide particles (Grilltex, $d_{\mathrm{p}} = 80$– 200 µm) are used as seeding particles. The idea of using the polyamide particles, which are normally used for surface coating, instead of more expensive fluorescent particles, is taken from a paper published by Keller et al. on the PIV measurement of large-scale vortices [Kel14]. The LED is used instead of a pulsed laser, due to safety concerns, as the setup could not be shielded completely against emitting light to the surroundings. The LED is mounted in front of one of the glass windows, emitting blue light with a wavelength between $\lambda = 454$– 462 nm. The light beam generated by the LED is focused into a narrow slit through two lenses, half-sphere and cylinder, placed between the LED and the access window. The camera (*pco.1600*, 14-bit CCD-camera with a *Micro-NIKKOR 105 mm f/2.8* lens) is mounted on a horizontal slide, within an endoscopic access pipe. The cameras field of view is redirected vertically upwards with an optical mirror through an acrylic glass-covered recess in the intermediate floor. With this setup, the velocities inside the vortex can be measured at different radii and heights above the pump intake, though the combination of radius and height are limited. A sketch of the setup can be seen in Figure 3-7. Since the LED beam fans out over the distance, the camera lens is set to a minimal focus depth by setting the aperture to *f/2.8* to minimize the error caused by the depth of field. The resulting pictures have a depth of field of about $t_f = 2$ cm. Camera and LED are synchronized by a microprocessor and controlled via a Laptop, running the microprocessor and camera control software. Double-image recordings are done at $f = 10$ Hz recording frequency, with a distance of $\Delta t = 1$ ms between the two images of a pair and an exposure time of $t = 0.1$ ms per picture. 600 double images are recorded for each operating point, which equals a measurement time of one minute each. The recordings are evaluated with the PIVView software from PIVTec, using a three step multi-grid refining with a starting grid size of 128x128 px^2, ending grid size of 64x64 px^2 and 50 % overlap. Mean velocities are determined by combining the results of 20 double images. Since a mirror must be used in the experiments to redirect the field of view, all images must be deskewed in a pre-processing step, using the PIVMap 3 software from PIVTec. The results for the second measurement set are published in the journal article "*Large Scale Experiments on the Formation of Surface Vortices with and without Vortex Suppression*" [Sze19].

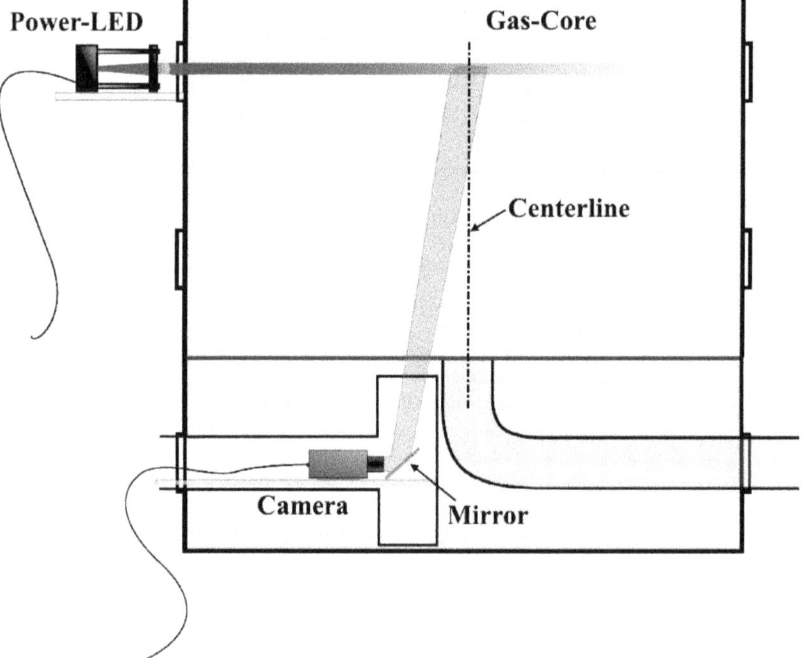

Figure 3-7: PIV measurement setup in the DN200 pilot plant.

3.2.3.3 DN15 Laboratory Plant

In the DN15 laboratory plant PIV measurements are conducted for a fixed submergence depth of $S^{\mathrm{LP}} = 0.11$ m ($S^* = 7.3$) and an open tube pump intake. The experiments are done for vertical inflow angles of $\beta = 0°, 15°, 30°$ and $45°$ at two different heights $h = 0.035$ m and 0.07 m above the pump intake. The measurements are conducted for the same four Froude numbers $Fr^{\mathrm{LP}} = 0.5, 0.7, 1.0$ and 1.4 as are investigated in the DN200 pilot plant. The transparent bottom is used in all experiments.

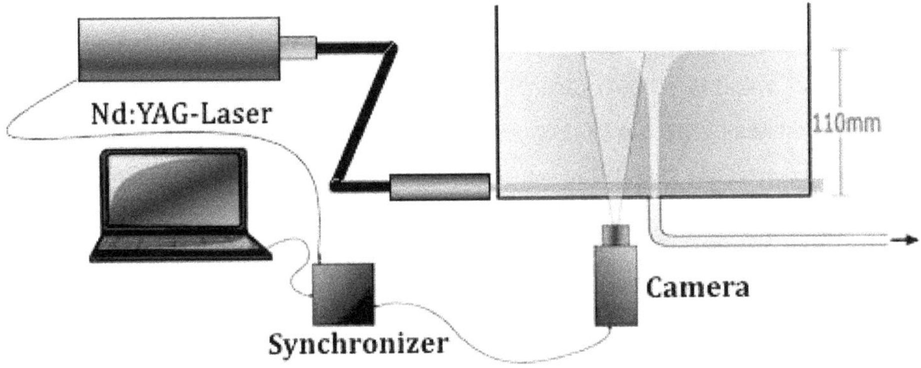

Figure 3-8: PIV measurement setup in the DN15 laboratory setup.

The PIV measurements in the DN15 laboratory setup are conducted with different measurement equipment than in the DN200 pilot plant setup: As a light source a 10 Hz double pulse laser (*Quantel, Nd:YAG*) is used, emitting light in the green spectrum with a wavelength of $\lambda = 532$ nm. A mirror equipped telescope arm connects the laser with a lens system. The lens system is used to spread the light into a *1 mm* thick horizontal sheet. For the recordings, the *pco.1600* CCD-camera is used, equipped with a *Zeiss Makro Planar 2/50 ZF.2* lens. The camera is mounted under the bottom of the vessel in a vertical angle (37° for $h = 0.035$ m and 58.5° for $h = 0.07$ m). This diagonal setup enables the recording of the vortex core region, without a disturbance of the view caused by the outlet pipe. As tracer particles acrylic glass encapsulated *Rhodamine B* particles with a size distribution of $d_{\mathrm{p}} = 20 - 50$ µm are used.

The measurements are conducted with a recording frequency of $f = 10$ Hz and a total amount of 500 double image pairs per measurement. Since the recordings are not orthogonal to the measuring plane, again a pre-processing is necessary to deskew the raw data, using the PIVMap3 software. After pre-processing is done, the pictures are evaluated using the PIVView2C with a three-step grid-refinement, this time with a larger grid: 160x160 px^2, 128x128 px^2 and 96x96 px^2 and with an overlay of 66 %.

The evaluated data is then further processed using a *Matlab* script to calculate the azimuthal and radial velocity as well as the vorticity and the circulation. Since the vortex center moves along the horizontal plane in a stochastic way, a moving grid is implemented to counter the vortex movements for the evaluation.

A single set of one-dimensional Laser Doppler Velocimetry (LDV) measurements is additionally conducted by the *ILA R&D GmbH* to provide a validation for the radial profiles of the azimuthal velocity. The LDV measurements are carried out for a submergence depth of $S = 0.11$ m, the Froude numbers of $Fr = 0.5$ and 1.0 and an inflow angle of $\beta = 45°$. The azimuthal profile is measured at heights of $h_{LDV} = 1$ mm, 35 mm and 70 mm above the pump intake. Pictures of the measurement setup can be seen in Figure 3-9.

Figure 3-9: Setup of the LDV measurements in the DN15 laboratory plant. Left: Side view of the LDV laser (bottom left side) and the experimental setup (right side). Right: Top view into the cylindrical vessel during the LDV measurements.

4 Results and Discussion

In this chapter the results of the different experiments are presented and discussed accordingly. The chapter starts with the presentation and discussion of the gas-core lengths measurements (see chapter 3.2), including the results of the dye experiments, followed by the presentation and discussion of the PIV measurements. After all results are presented and discussed, the experimental setup and measurements are evaluated by an error estimation. The last part of this chapter covers a compilation of design recommendations and a proposition for the improvement of the Burgers-Rott vortex model.

4.1 Measured Gas-Core Lengths

In this part, the vortex formation is described qualitatively and quantitatively. Beginning with the vortex formation in both scales, without the implementation of vortex breakers. After that, the influence of the inflow angle and the efficiency of the different vortex breaker designs are shown and evaluated.

4.1.1 Gas-Core Formation

The vortex formation can be seen exemplarily in Figure 4-1, showing photographs of different gas-core lengths for a dimensionless submergence depth of $S^* = 7.3$. The symmetric inflow conditions lead to a strong and stable vortex formation in all experiments with an inflow angle of $\beta = 45°$, regardless of the setup size. The resulting vortices are quasi stationary with only minor fluctuations in the tip length of the forming gas-cores.

In both plant sizes the gas-core development shows a similar behavior. Looking at the results for the open pump intakes, plotted in Figure 4-3 for all three different submergence depths, it is visible that the development of the dimensionless gas-core lengths in dependency of the Froude number is sigmoid (S-shaped). In general, the curves for the DN15 laboratory plant are steeper than the curves for the DN200 pilot plant, although the difference is decreasing with increasing submergence depth.

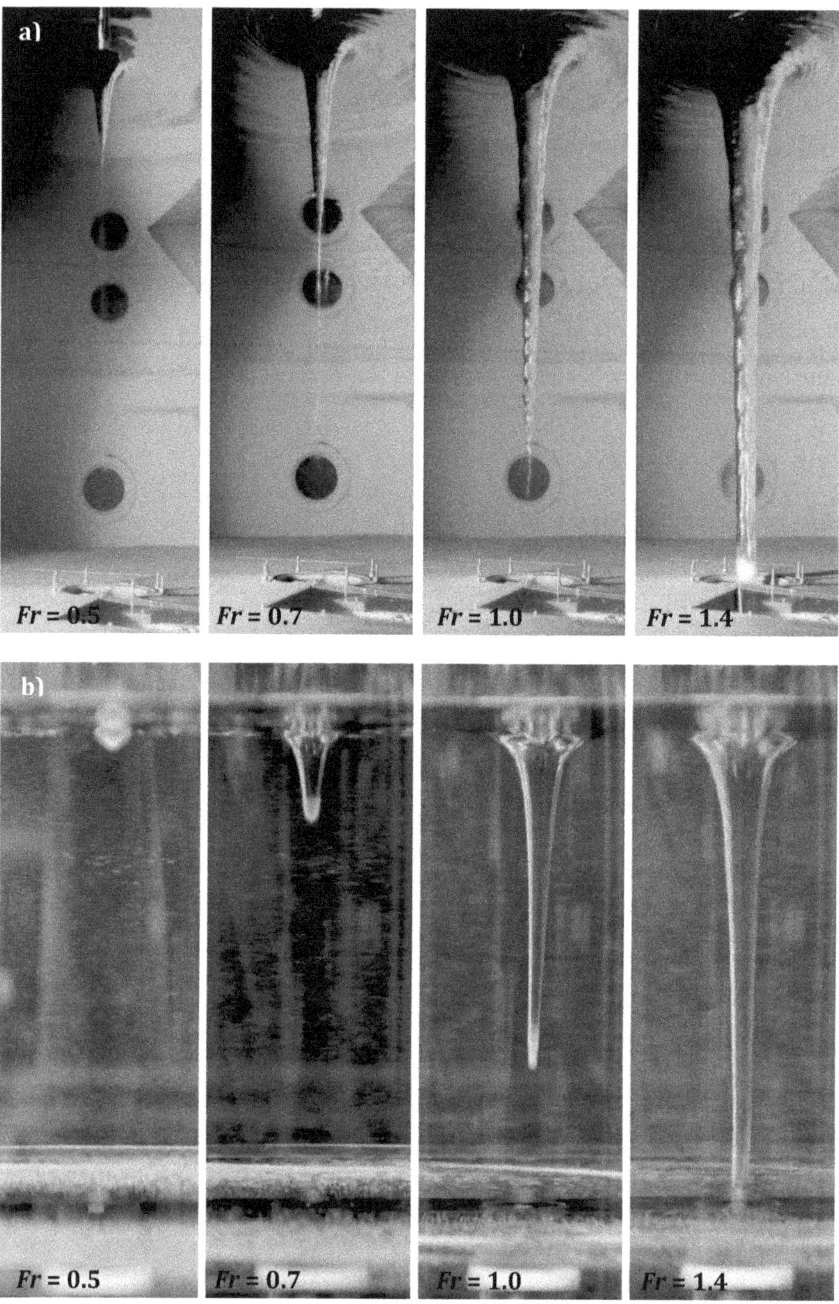

Figure 4-1: Photographs of vortex-core lengths for an open tube pump intake, $S^* = 7.3$ and $\beta = 45°$ at different Froude numbers. a) DN200 Pilot Plant with intermediate floor, $S = 1.46$ m; b) DN15 Laboratory Plant, $S = 0.11$ m.

Minimizing the numbers of variables needed to describe the curve, it is assumed that the circulation is mostly depending on the Froude number, thus eliminating the circulation from the equation. The validity of the assumption can be confirmed by the results of the PIV measurements, as can be seen in Figure 4-2, depicting the linear dependency of the bulk circulation on the Froude number. The development of the gas-core length can therefore be described by a sigmoid equation, of the form

$$L^{*} = tanh\frac{C \cdot Fr^{\alpha}}{S} \tag{4-1}$$

Which is only depending on the Froude number Fr and the submergence depth S as well as the fitting constant C and the fitting exponent α. For the measurements in the DN200 pilot plant with an open pump intake, the fitting constant C is 1 and the exponent α has a constant value of 2.8, showing the neglectable differences in the circulations between the varying submergence depths in the large-scale experiments. The influence of the intermediate floor in the DN200 pilot plant is also shown to have a minor negative impact on the critical Froude number Fr_{crit}, as can be seen in Figure 4-3b (colored diamonds). Though for moderate Froude numbers, the influence of the intermediate floor is larger. This is most likely caused by the higher symmetry of the system, when using the intermediate floor. For the DN15 laboratory plant, the exponent α has a constant value of 5.6, while the fitting constant changes with each submergence depth: $C_{S^*=5.3} = 0.45$, $C_{S^*=7.3} = 0.15$, $C_{S^*=10.7} = 0.05$.

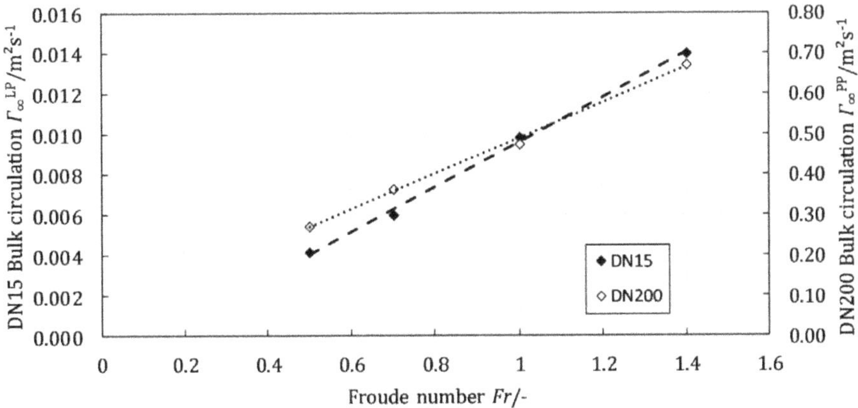

Figure 4-2: Linear dependency of the bulk circulation on the Froude number shown for both experimental sizes.

The difference in the exponent and constant values between the two setup sizes can be explained by scaling differences and the fact that surface tension and viscosity effects might have an influence in the laboratory scale. The exponent in the laboratory scale is twice as large as the exponent in the pilot scale for all submergence depths, due to the steeper gain of the gas-core lengths in the laboratory scale. The values for the fitting constant C also change between the different submergence depths in the laboratory scale (unlike in the pilot scale, where $C = 1$ for all results), possibly due to viscosity and surface tension effects on the vortex development. The curves of the sigmoid functions are also plotted in Figure 4-3 for the open pump intakes and show a good agreement with the experimental data.

Rearranging eq. 4-1 allows to calculate the Froude number for each submergence depth and gas-core length

$$Fr = \left(\frac{S}{C} \cdot tanh^{-1}(L^*)\right)^{\frac{1}{\alpha}} \qquad (4\text{-}2).$$

This further opens the possibility to calculate the critical Froude number, by setting the gas-core length L^* to 1. Though it is better to use a defined cutoff value of 0.9 for L^*, as 1 is the upper asymptotic border for the sigmoid function and the deviation between the function and the experimental values increases for $L^* \rightarrow 1$. This also imposes an additional safety margin for the reliable operation of the pump system. With a constant value of $L^* = 0.9$, equation 4-1 can be simplified to

$$Fr_{\text{crit,90\%}} = \left(\frac{1.47 \cdot S}{C}\right)^{\frac{1}{\alpha}} \qquad (4\text{-}3)$$

or in case for the experiments conducted for an open pump intake in the DN200 pilot plant

$$Fr_{\text{crit,90\%}}^{\text{PP}} = (1.47 \cdot S)^{0.36} \qquad (4\text{-}4).$$

The calculated critical Froude numbers are $Fr_{\text{crit,90\%}}(S = 1.08\text{ m}) = 1.2$, $Fr_{\text{crit,90\%}}(S = 1.46\text{ m}) = 1.3$ and $Fr_{\text{crit,90\%}}(S = 2.13\text{ m}) = 1.5$. They are in good agreement to the measured Froude numbers $Fr_{\text{crit,90\%}}^{\text{exp}}(S = 1.46\text{ m}) = 1.4$ and $Fr_{\text{crit,90\%}}^{\text{exp}}(S = 2.13\text{ m}) = 1.5$. Since the largest measured gas-core length for $S = 1.1$ m is $L^* = 0.84$ at $Fr = 1.10$, which is less than the required 90 %, the results for this specific submergence depth are therefore excluded from the comparison. [Sze19]

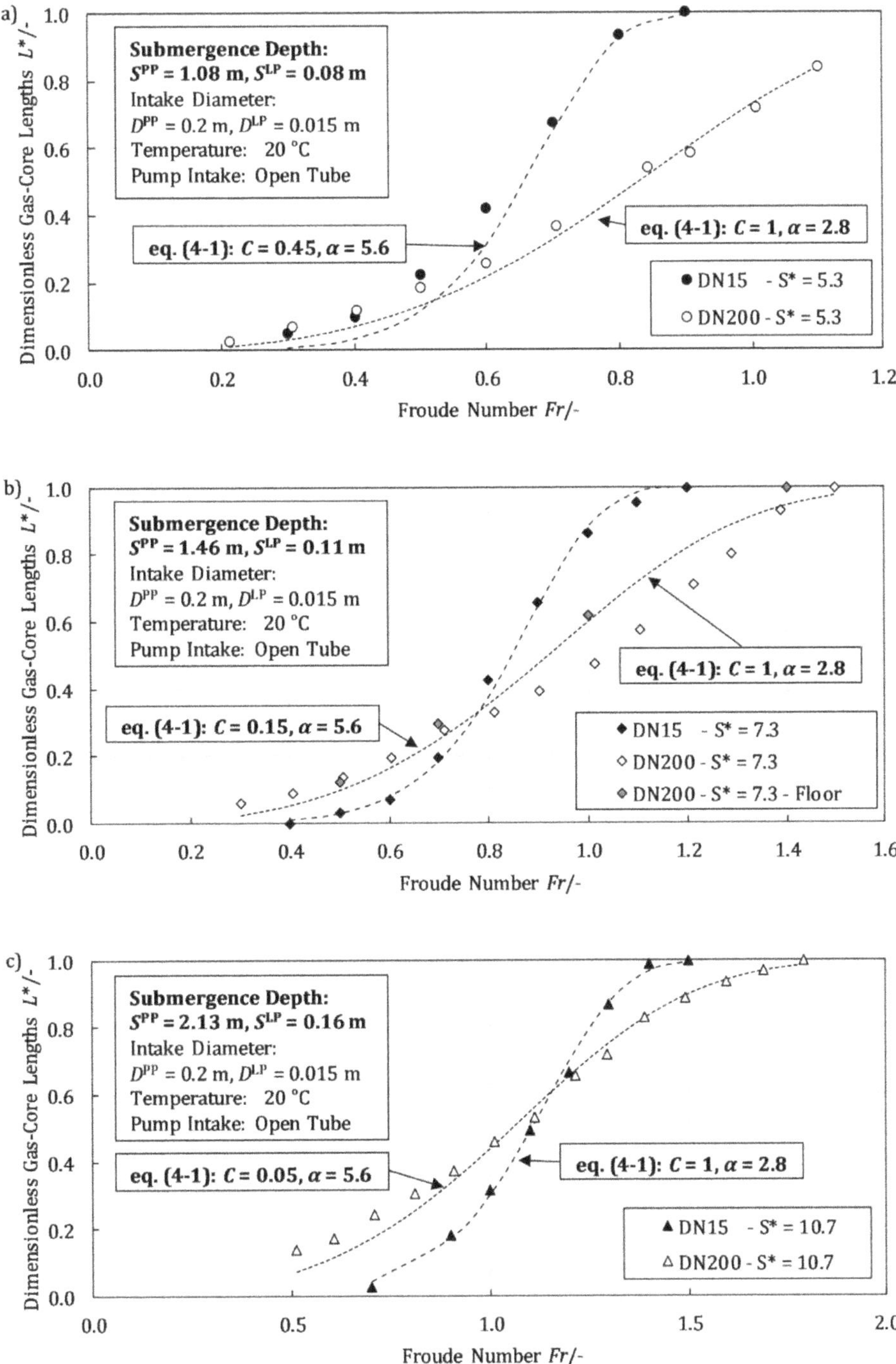

Figure 4-3: Development of the gas-core lengths for an open pump intake in dependency of the Froude number in the DN15 and the DN200 plants. a) $S^* = 5.3$, b) $S^* = 7.3$ and c) $S^* = 10.7$.

4.1.2 Influence of the Induced Momentum on the Gas-Core Formation

The experiments conducted in the DN15 laboratory plant show that with an increasing inflow angle, stronger and more stable vortices occur. The strongest observed vortices occur in experiments with this inflow angle of $\beta = 45°$, since this is the largest investigated angle in both setups. Further test measurements with horizontally bend inflow pipes ($\beta = 90°$) show an even stronger effect on the vortex formation, but as they could not be reproduced in the DN200 pilot plant setup due to constructive limitations, no recordings are made. Although investigated, no gas-core lengths can be measured for experiments with a directly vertical inflow ($\beta = 0°$), as no vortex formation is observable within the limits of the experiment. For smaller inflow angles ($\beta = 15°$ and $30°$), the vortex formation is more unstable, especially for larger Froude numbers.

Due to the smaller momentum induced into the system, the vortex formation also occurs at higher Froude numbers and the critical value is reached at even higher Froude numbers. This can be seen in Figure 4-4 for the experiments conducted in the DN15 laboratory plant. Not only the length, but also the shape of the gas-core changes with smaller inflow angles. The smaller the inflow angle, the thinner the gas-core, as can be seen in Figure 4-6a & b. This is due to two causes: First, with a decrease in the inflow angle, the tangential momentum decreases, reducing the overall circulation in the system and increases the influence of the surface tension and viscosity on the vortex development. The second cause is that the exit stream from the inflow pipes is deflected from the bottom of the vessel, causing more turbulence in the experiments with smaller inflow angles.

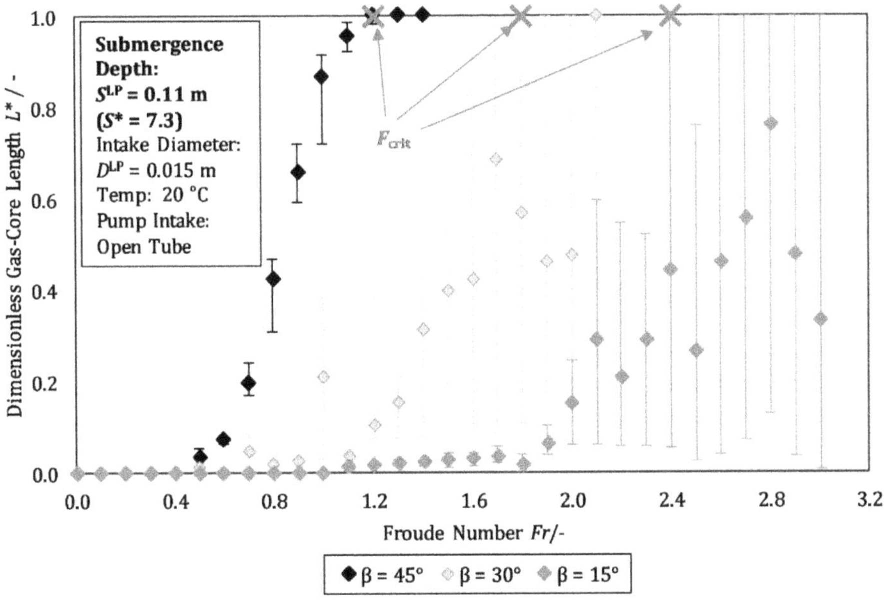

Figure 4-4: Influence of the inflow angle β in the DN15 laboratory plant. Plotted against the Froude number are the averaged dimensionless gas-core lengths and their maximum deviation. Marked in red are the critical Froude numbers for each inflow angle.

In the DN200 pilot plant the influence of the vertical inflow angle on the vortex stability and gas-core shape is also visible, see Figure 4-6c & d. Contrary to the laboratory scale experiments, the average gas-core length does not change by much between $\beta = 15°$ and $\beta = 45°$. A possible reason is the huge pressure loss above the pump intake and the neglectable influence of the viscous and surface tension forces in comparison. Nonetheless, the gas-core length is subjected to heavy fluctuations, as can be seen in Figure 4-5. In general, a minimizing the inflow angle is preferable for the operation of a pump system, as it reduces the amount of induced momentum and therefore the vortex development. Additionally, even when the gas-core reaches the pump intake and gas is entrained into the system, overall the gas-entrainment is smaller when compared to experiments with larger inflow angles, as the gas core is noticeably narrower. A disadvantage of a smaller induced momentum is the less predictable vortex behavior, which can be seen in Figure 4-5, where the average gas-core length for the DN200 pilot plant is more or less the same for both inflow angles, but the maximum gas-core length is much higher for the smaller inflow angle ($\beta^{PP} = 15°$) and almost reaches into the pump intake already at $Fr = 0.9$.

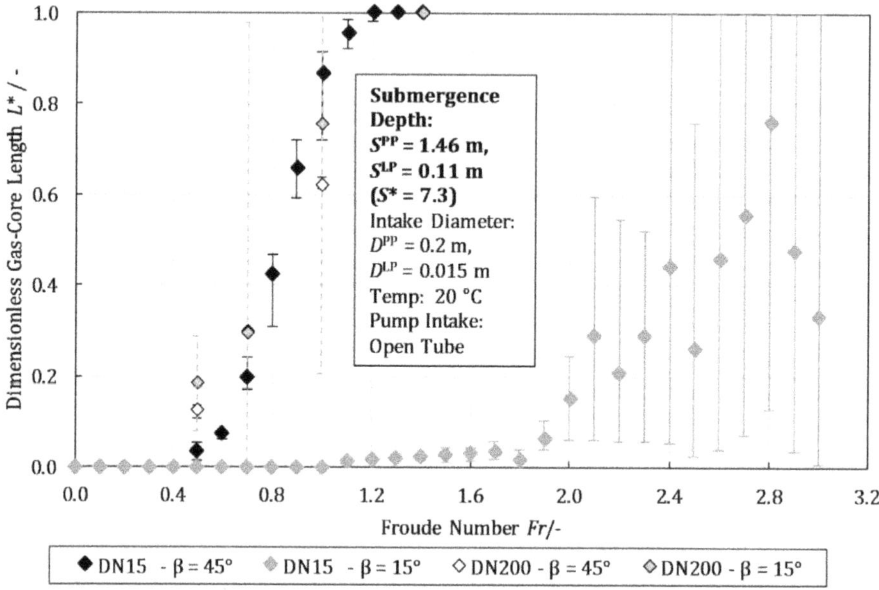

Figure 4-5: Comparison of the inflow angle β between the setup sizes. Plotted against the Froude number are the averaged dimensionless gas-core lengths and their maximum deviation. Filled markers denote the results from the DN15 laboratory plant, while hollow markers indicate results from the DN200 pilot plant.

4.1.3 Scale-up Comparison

Comparing the two plant sizes has proven to be difficult, due to the fact that not all scaling criteria (see chapter 2.4) are met. Nonetheless, a similar behavior in the vortex development and the lengths scales for the gas-cores can be found, when the inflow momentum is large ($\beta = 45°$). For smaller inflow angles, the behavior between the two plant sizes diverges greatly, as the viscous and surface tension force effects differ between the sizes. In the DN15 laboratory plant, the formation of the gas-core is delayed by the viscosity and surface tension. In the DN200 pilot plant the gas-core shape and stability are affected by the reduced momentum, but the mean gas-core lengths are not.

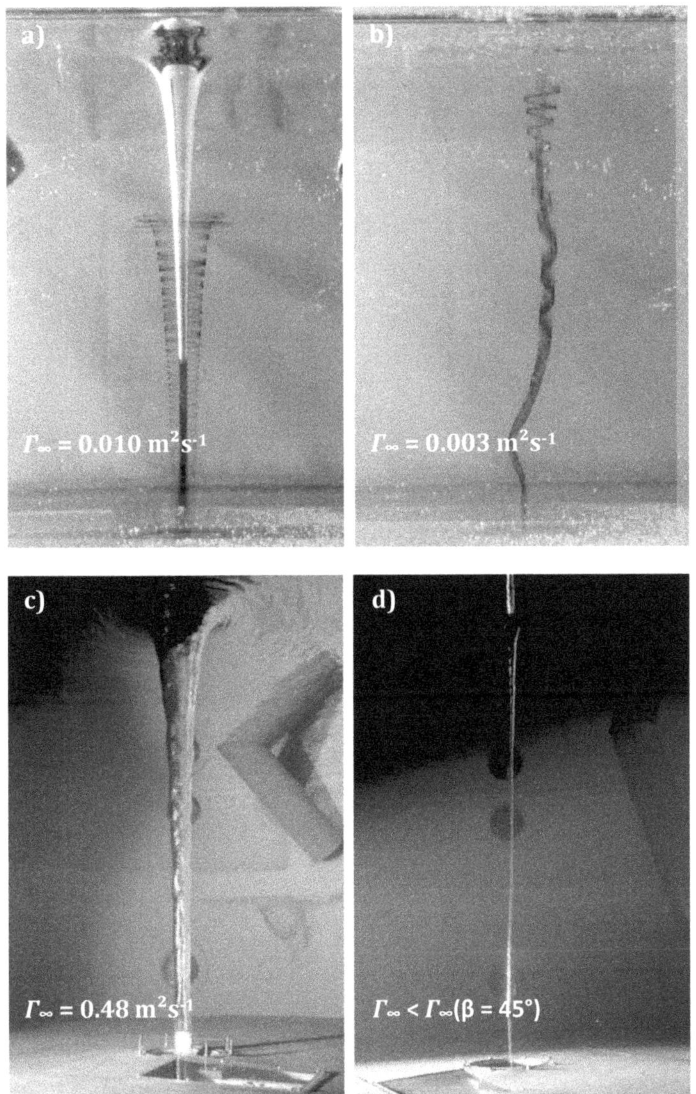

Figure 4-6: Photographs of the gas-core vortices for different inflow angles. Top: DN15 laboratory plant at $Fr = 1.0$, vortex flow visualized by ink (dark grey); $\beta = 45°$ (a) and $\beta = 15°$ (b). Bottom: DN200 pilot plant at $Fr = 1.4$; $\beta = 45°$ (c) and $\beta = 15°$ (d).

4.1.4 Vortex Prevention Measures

4.1.4.1 Influence of the Pump Inlet Shape on the Vortex Formation

The investigated intake shapes are found to have only minor influence on the vortex formation and gas-core length. In Figure 4-7 the results of the intake shape variations in the DN15 laboratory plant are shown. As in previous figures, the dimensionless gas-core lengths are plotted against the Froude number. The Figure shows that the use of the funnel shape results in a smaller gas-core at the same Froude number, as compared to the open tube shape, but the difference is generally very small. When using the sieve or the recess shape, the resulting gas-cores are a bit longer than for the open tube shape. As none of the intake structures has a significant effect on the vortex formation, the difference in gas-core length can be explained by looking at the intake area of each structure, see Table 3-3.

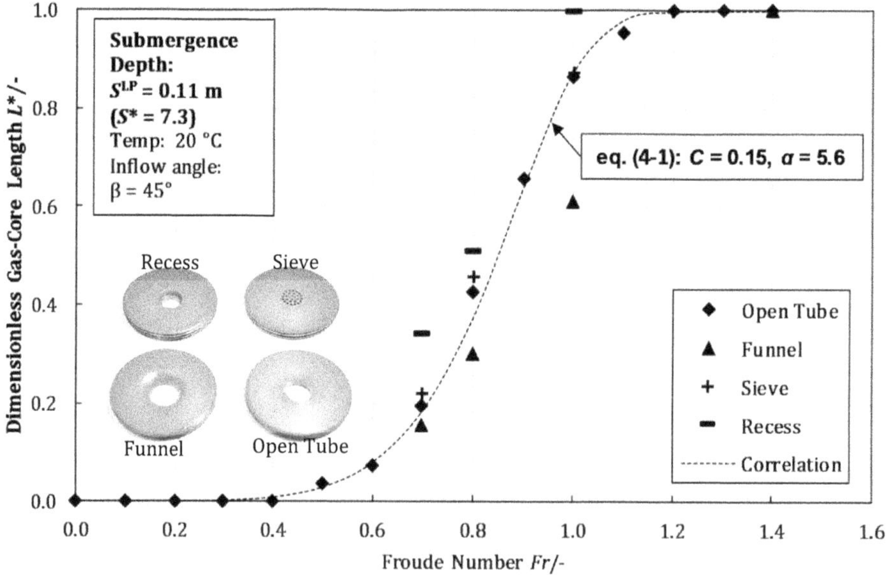

Figure 4-7: Gas-core lengths for different intake shapes in the DN15 plant.

Since the intake areas of the recess and the sieve are smaller than for the open tube, for the same volume flow rates the velocity at the intake are higher. The higher velocities in an overall smaller intake area increase the pressure drop at the surface and thereby create longer gas-cores.
In contrast, the gas-core lengths measured for the funnel intake shape are slightly shorter at the same volume flow rate. If the reduced gas-core length is due to the improved inflow conditions of the funnel or because of the increased intake area created by the wider intake diameter cannot be distinguished by the measurements.

The experiments with the different intake shapes in the DN200 pilot plant show a similar development of the gas-core lengths, as can be seen Figure 4-8. Also, in this case the influence of the intake shapes on the gas-core length development is either neglectable (funnel and 45°-bend) or even negative (sieve), when the intake diameter is reduced. These experiments confirm the general knowledge that a reduction of the intake area leads to an increase of the intake velocity and therefore a stronger vortex formation with longer gas-cores. However the experiments also show that a simple increase of the intake area through the implementation of a funnel doesn't lead to a decrease in the gas-core length, as proposed by the American National Standard, see chapter 2.3.3 or [Hyd12]. The reason why an increase in the intake area doesn't lead to a reduced length of the gas-core is that the critical diameter is the narrowest diameter in the pump intake. As the pipe behind the intake, which leads to the pump, is of constant diameter in all experiments, increasing the intake area above the pipe proves inefficient. This leads to the conclusion that in the ANSI correlation of the Hydraulic Institute (eq. 2-47), for the intake dimeter D_{Intake} the smallest diameter in the intake pipe must be chosen.

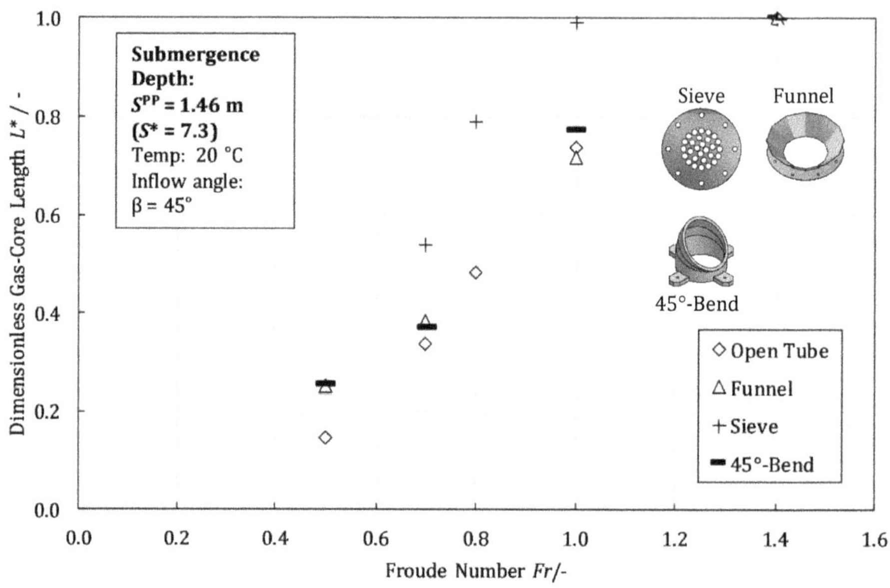

Figure 4-8: Gas-core lengths for different intake shapes in the DN200 plant.

4.1.4.2 Vortex Breakers

The experiments regarding the effectiveness of the designed vortex suppression structures are presented in the following. First for the DN15 laboratory plant and then for the DN200 pilot plant.

DN15 Laboratory Plant

Experiments are carried out for each structure at the maximum inflow angle of $\beta = 45°$ and for all three submergence depths, with the exception of the Inner-Cross and the variation of the wide Baffle-Plate with larger baffles, which are only investigated at the lowest submergence depth ($S^* = 5.3$). In all experiments, the gas-core lengths are measured in steps of $\Delta Fr = 0.5$, until the critical Froude number $Fr(L^* = 1)$ or, due to limitations of the experimental setup, the maximal possible Froude number of $Fr = 6.5$, is reached. During the execution of the experiments it has been observed that as soon as the gas-core reaches the vortex breaker, gas starts to be entrained. Since the critical inflow conditions are thereby met, the gas-core length is counted in this case as $L^* = 1$ and the experiments are stopped. All measured gas-core lengths are plotted in dimensionless form in dependency of the Froude number in Figure 4-9. All vortex breakers, except for the Inner-Cross shape, enable a much higher volume flow rate/Froude number, before the critical gas-core length is reached. For several structures and with increasing submergence depth, the setup can be operated on its maximum capacity of $Fr = 6.5$ without reaching critical inflow conditions, that is gas-entrainment.

The results of the gas-core lengths measurements can be broadly separated into different categories, which define the efficiency of the tested structures on the reduction of the gas-core length. The results show that the efficiency is mostly depending on the diameter and, to a lesser degree, on the type of the vortex breaker. All experiments confirm that wide vortex breakers with $d = 4D$ are more efficient than narrower structures with $d = 2D$. Additionally, for vortex breakers of the same diameter, structures with a plate are more efficient than structures without one. The variation in height of a structure shows more ambiguous results. For medium and high submergence depths, the tall cross structure reduces the gas-core length more effectively when compared to the cross structure, as can be seen in Figure 4-9b & c. But when the submergence depth is reduced, as can be seen in Figure 4-9a, the height has no positive effect anymore.

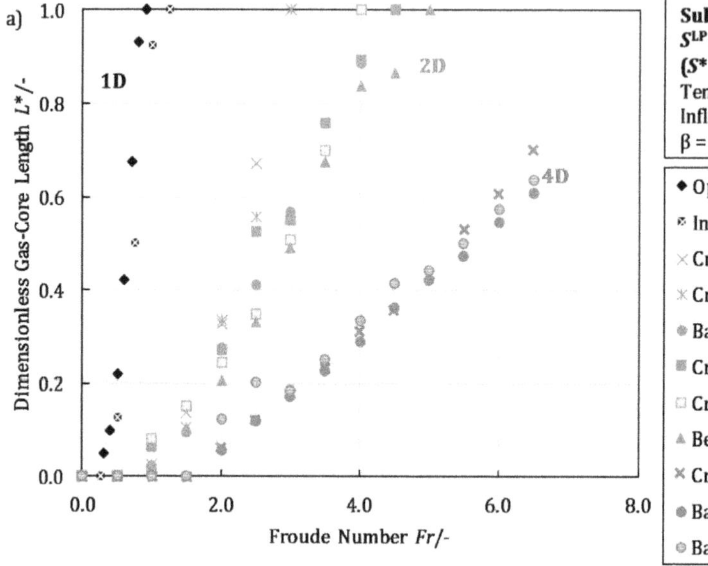
a)
1.0
0.8
0.6
0.4
0.2
0.0
0.0
2.0
4.0
6.0
8.0
Dimensionless Gas-Core Length L*/-
Froude Number Fr/-
1D
2D
4D

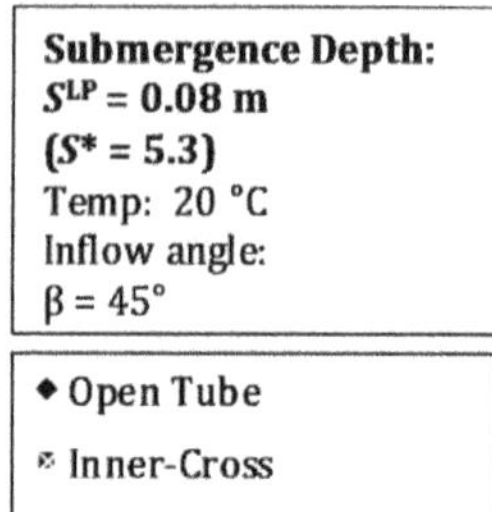
Submergence Depth:
S^LP = 0.08 m
(S* = 5.3)
Temp: 20 °C
Inflow angle:
β = 45°

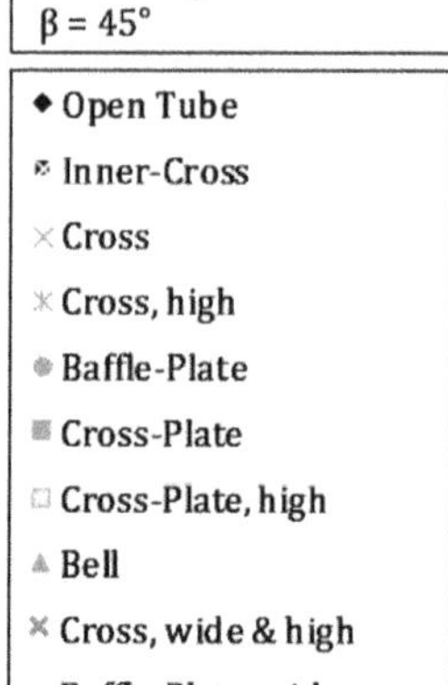
Open Tube
Inner-Cross
Cross
Cross, high
Baffle-Plate
Cross-Plate
Cross-Plate, high
Bell
Cross, wide & high
Baffle-Plate, wide
Baffle-Plate, wide (var.)

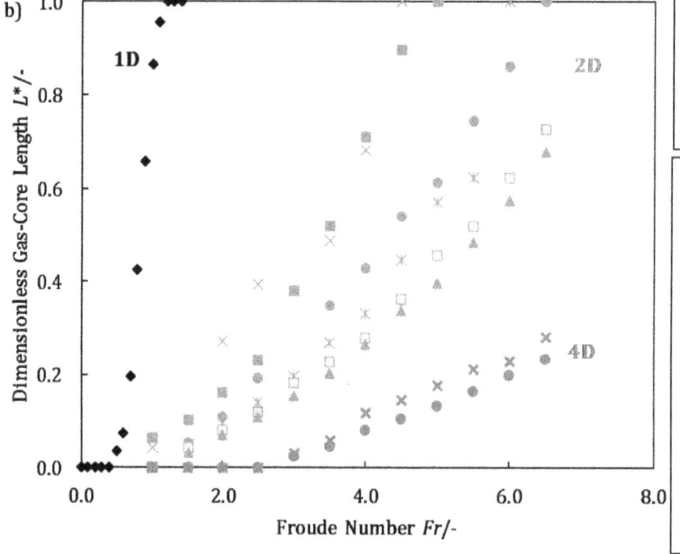
b)
1.0
0.8
0.6
0.4
0.2
0.0
0.0
2.0
4.0
6.0
8.0
Dimensionless Gas-Core Length L*/-
Froude Number Fr/-
1D
2D
4D
Open Tube
Cross
Cross, high
Baffle-Plate
Cross-Plate
Cross-Plate, high
Bell
Cross, wide & high
Baffle-Plate, wide

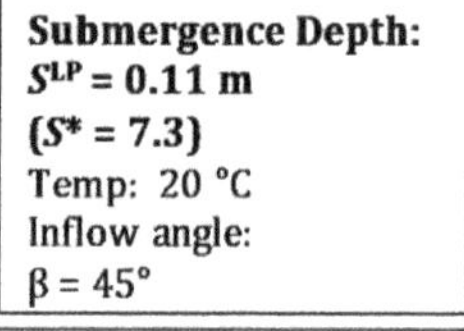
Submergence Depth:
S^LP = 0.11 m
(S* = 7.3)
Temp: 20 °C
Inflow angle:
β = 45°

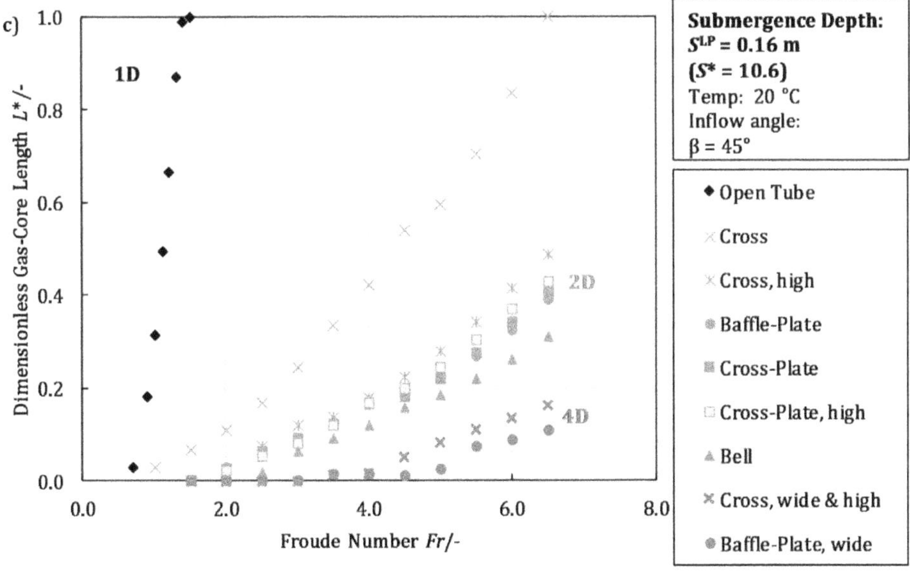

Figure 4-9: Results of the vortex breaker measurements in the DN15 plant. Plotted are the dimensionless gas-core lengths against the Froude number. a) For the low submergence depth of S = 0.08 m, b) for the medium submergence depth of S = 0.11 m and c) for the highest submergence of S = 0.16 m. Pictures of the vortex breakers can be seen in Figure 3-6.

An assumption for this observation is that the taller vortex breakers are more efficient in breaking the vortex, as they reach further into the liquid phase, but therefore, since they are taller, the gas-core also doesn't have to be as long when it reaches the vortex breaker and causes gas-entrainment. Comparing the results of the Bell and Baffle-Plate vortex breakers shows that it is more efficient to add a rim onto the plate instead of increasing the height of the structure. Since diameter and plate are found to be the most important features of the vortex breakers, it is no surprise that the difference between the two variations of the wide Baffle-Plate structures are negligible small. Like the height of the Cross-structures, also the difference between Cross-Plated and Baffle-Plated vortex breakers show inconclusive results. For low and high submergence depths the results of both structures are almost identical, but for the medium submergence depth, the Baffle-Plate shows better results. At last, as mentioned above, it is visible from the results that the implementation of the Inner-Cross structure does not result in a significant increase in the critical Froude number.

The effect of the vortex breakers on the flow structures in the liquid phase and therefore on the vortex development can also be explained by comparing the recordings of the dye experiments for the different structures, as are shown in Figure 4-10.

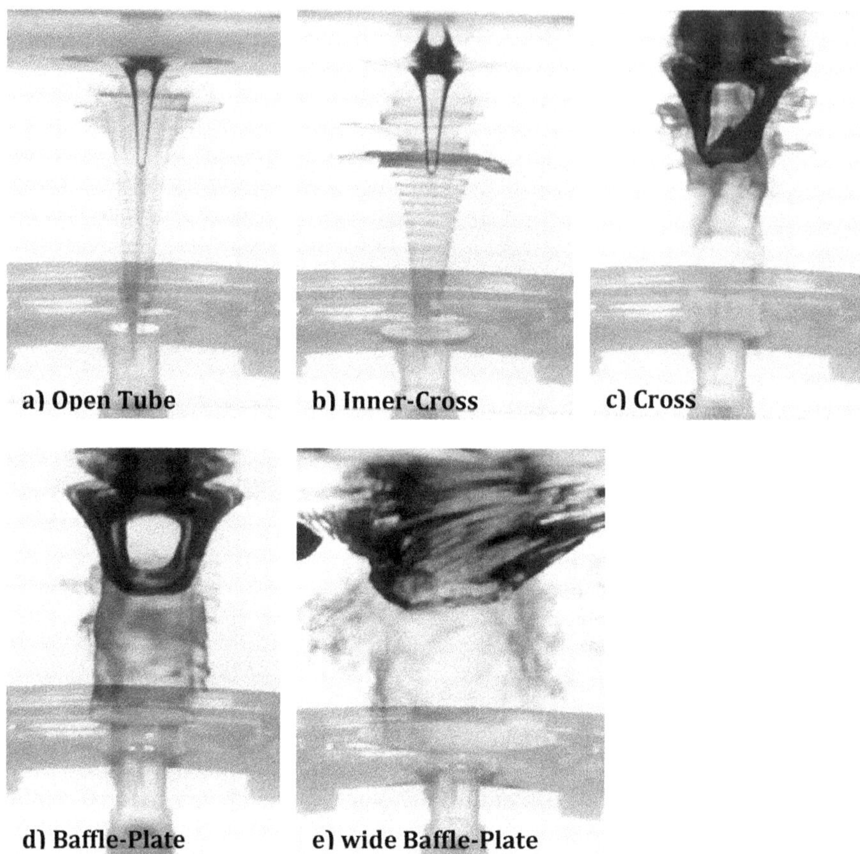

Figure 4-10: Flow structures above the pump intake in the DN15 laboratory plant visualized by ink for $Fr_{0.5}$. a) Open Tube, b) Inner-Cross, c) Cross, d) Baffle-Plate & e) wide Baffle-Plate vortex breaker.

Comparing the flow structures of the open tube (Figure 4-10a) and Inner-Cross structure (Figure 4-10b) with flow fields of the other structures shows that for an effective vortex prevention the vortex breaker has to protrude from the pump intake into the liquid, as a vortex prevention inside the intake pipe isn't effective. Comparing the Cross (Figure 4-10c) with the Baffle-Plate (Figure 4-10d) shows the positive effect of the Plate on the vortex suppression. While the cross disturbs the formation of the vortex core, liquid (and later gas) can still flow into the intake vertically. The plate, especially in combination with a cross or baffles underneath, not only interferes with the vortex development, but also prevents a direct, vertical flow from the surface into the pump intake. A wider plate enhances this effect, leading to a much broader, but shallower gas-core. Most likely due to a reduction of the maximal pressure loss on the liquid surface by distributing the overall pressure loss over a wider area.

DN200 Pilot Plant

Also in the DN200 pilot plant dye experiments are conducted, coloring the vortex core region with blue ink, as can be seen in Figure 4-11 for the open pump intake and the cross-plate vortex breaker. While the vortex core region for the open pump intake is roughly $1/5^{th}$ of the diameter of the pump intake wide, the vortex core diameter increases to the diameter of the baffle, when using a cross-plate vortex breaker, see Table 4-1. As a result, the pressure drop, causing the gas-core to form, is distributed over an area which is 100 times larger, again leading to a wider but much shorter gas-core. At the same time the plate obstructs the strait way from the surface to the intake, resulting in a more radial intake of water instead of vertical, further reducing the vortex strength.

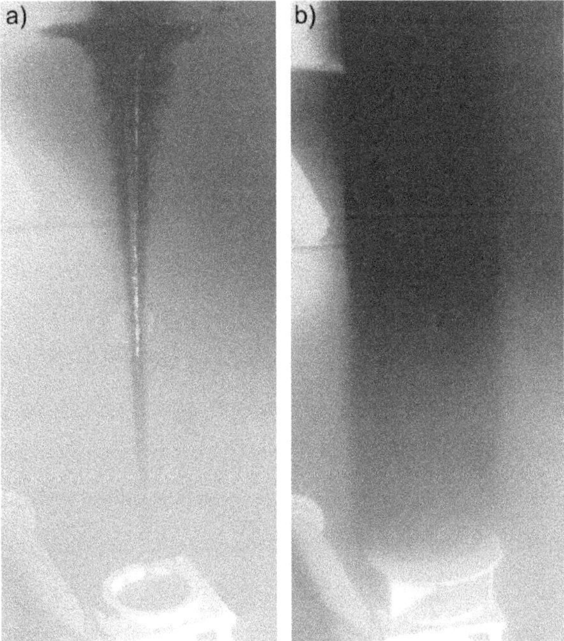

Figure 4-11: Visualization of the vortex-core region in the DN200 pilot plant with blue ink. a) For the open pump intake and b) the cross-plate vortex breaker.

Table 4-1: Vortex-core radius to intake diameter ratios for different vortex breaker, based on the dye experiments.

Vortex Breaker:	**Vortex-core radius to intake diameter ratio** $r_0 \cdot D^{-1}/-$	
	DN200 Pilot Plant	**DN15 Laboratory Plant**
Open Tube	0.2	0.2
Inner-Cross	-	0.2
Cross	1.7 – 2	1.7 – 2
Baffle-Plate	-	2
Baffle-Plate, wide	-	4
Cross-Plate	2	2
Cross-Plate, wide	4	4

Figure 4-12 shows the results for the measurements with the vortex breakers in the DN200 pilot plant for the low and medium submergence depths. By design, a volume flow rate of Q = 755 m^3h^{-1}, resulting in a Froude number of Fr = 4.8, is the highest volume flow rate at which the plant can be safely operated. Therefore, no critical submergence is measurable for the investigated vortex breakers, as the critical Froude number for them is considerably higher than 4.8. Like in the laboratory sized experiments, the cross, cross-plate and baffle-plate devices are found to be very efficient vortex breakers, preventing gas entrainment even for large Froude numbers. The wide cross-plate structure with a diameter of 4D, is only shown for the lowest submergence depth, as this structure prevents the gas-core formation so efficient, that only at this submergence depth a gas-core could be observed before the maximum volume flow rate is reached. For the cross, cross-plate and baffle-plate vortex breakers, no gas-entrainment is observed during all experiments, while experiments without vortex suppression are limited to volume flow rates of Q < 160 m^3h^{-1}, to avoid gas-entrainment.

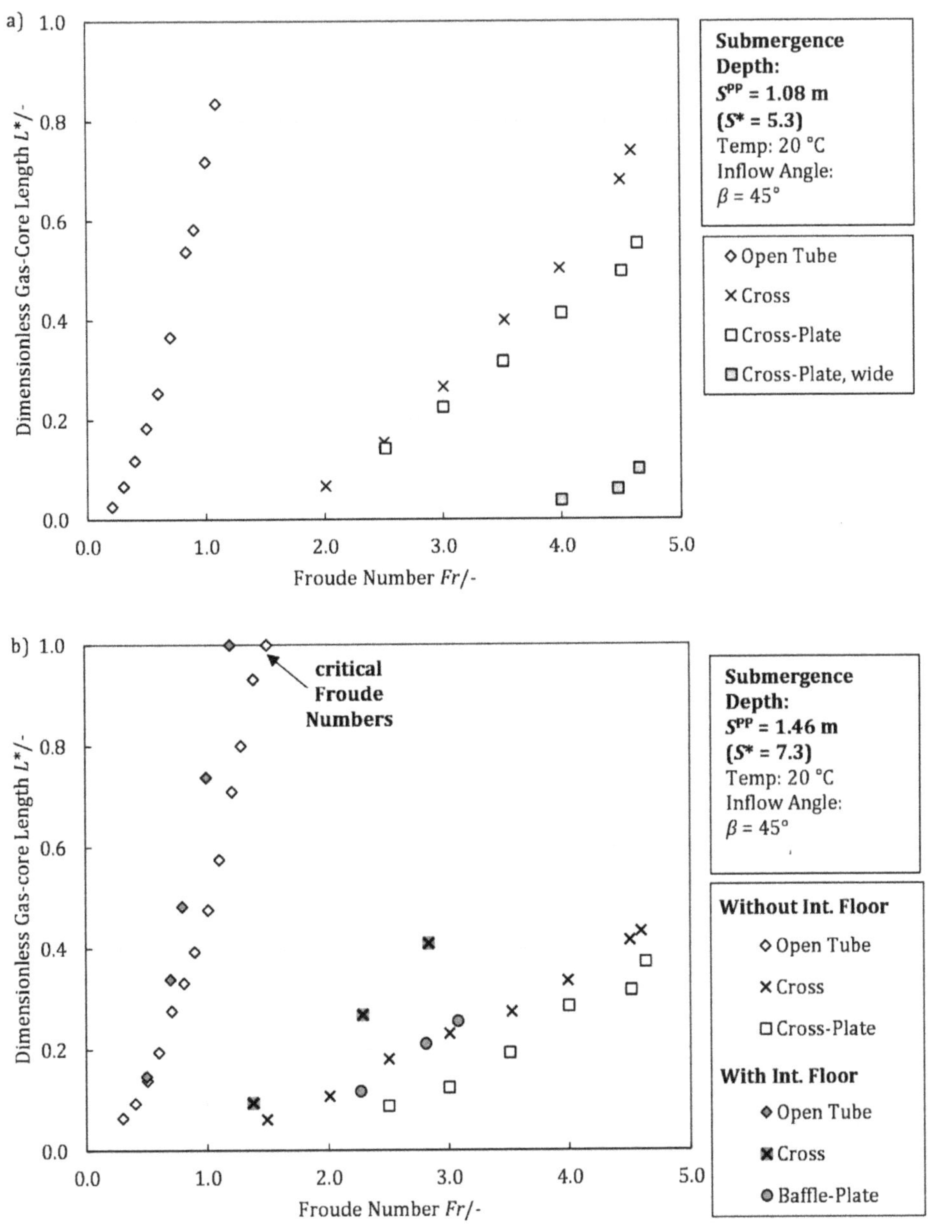

Figure 4-12: Results of the vortex breaker measurements in the DN200 pilot plant. Plotted are the dimensionless gas-core lengths against the Froude number. a) For the low submergence depth of S = 1.08 m & b) for the medium submergence depth of S = 1.46 m. Marked in red are the experiments conducted with the intermediate floor installed at the height of the pump intake. Pictures of the vortex breakers can be seen in Figure 3-3.

Error Analysis

In the DN15 laboratory setup, the maximum deviation from the mean gas-core length is 6.4 % for measurements with an open pump intake and an inflow angle of 45°, while most experiments show fluctuations below 2.3 %. For measurements with a smaller inflow angle, the decreased circulation and the higher turbulence in the system leads to a more stochastic vortex development, with fluctuations in the gas-core length of $L^* = 0$ to $L^* = 1$ within a few seconds, which makes the use of mean values questionable.

The maximum measurement uncertainty for the measurements with vortex breakers caused by fluctuations of the gas-core length is 3.9 %, while most experiments have deviations from the mean, which are below 1 %. Additionally, it is observed in measurements with cross shaped breakers, that the gas-core tip is lacerated at higher Froude numbers, which complicates the determination of the tip end slightly.

Same observations can be made for the DN200 pilot plant, where the maximum deviation from the mean gas-core length is 6.4 % for measurements with an open pump intake and an inflow angle of $\beta = 45°$.

The biggest source for errors in the experiments are accumulating gas pockets in the inflow pipes, which lead to an uneven inflow and therefore changed circulation within the cylindrical vessel. The error is easier to detect in the DN200 pilot plant, as each inflow pipe is equipped with its own flow meter, but harder to correct, since the evacuation of the gas-pocket is more cumbersome. Other sources of error are minor deviations in the submergence depths between the experiments, although rigorous metering has been done to minimize the error. To avoid dynamic errors, a waiting time is implemented before each measurement, to ensure steady state conditions are reached. Lastly, the error caused by the automatic evaluation is minimized by using a large sampling size and making comparative evaluations by hand.

4.1.5 Comparison with Literature Data

In this part, the experimentally determined gas-core lengths are compared with literature data and the correlation for the critical submergence from the *Hydraulic Institute* (see chapter 2.3.3).

To compare the results with the ANSI correlation, eq. 2-46 is changed to

$$Fr_{\mathrm{crit}} = \frac{\frac{S}{D_{\mathrm{Intake}}} - 1}{2.3} \quad (4\text{-}5),$$

which enables the direct comparison with the experimental data for a fixed submergence depth. The results for both plant sizes and all submergence depths are written in Table 4-2. Comparing the experimental values for the critical Froude number with the ones calculated with eq. 4-5 shows that the critical Froude numbers derived by the correlation overestimate the measured critical values by a large margin.

The reason for this difference is most likely that the circulation in the experimental setup is above the threshold of the ANSI correlation and that therefore the correlation is not applicable, although the volume flow and intake velocities in the setup are within the given boundaries of the correlation.

Table 4-2: Experimentally determined critical Froude numbers and theoretical critical Froude numbers for the DN200 pilot plant and the DN15 laboratory plant setup with β = 45°.

Parameter:	**DN200 Pilot Plant**			**DN15 Laboratory Plant**		
Submergence depth S/m	1.07	1.46	2.13	0.08	0.11	0.16
Critical Froude number, experiment $Fr_{\mathrm{crit}}^{\mathrm{EXP}}$/-	1.2†	1.5	1.8	0.9	1.2	1.5
Critical Froude number, correlation $Fr_{\mathrm{crit}}^{\mathrm{ANSI}}$/-	1.9	2.7	4.3	1.9	2.8	4.2

†: Eq. 4-2 is used to estimate the critical Froude number for S = 1.07 m in the DN200 pilot plant with a value of L^* = 0.9.

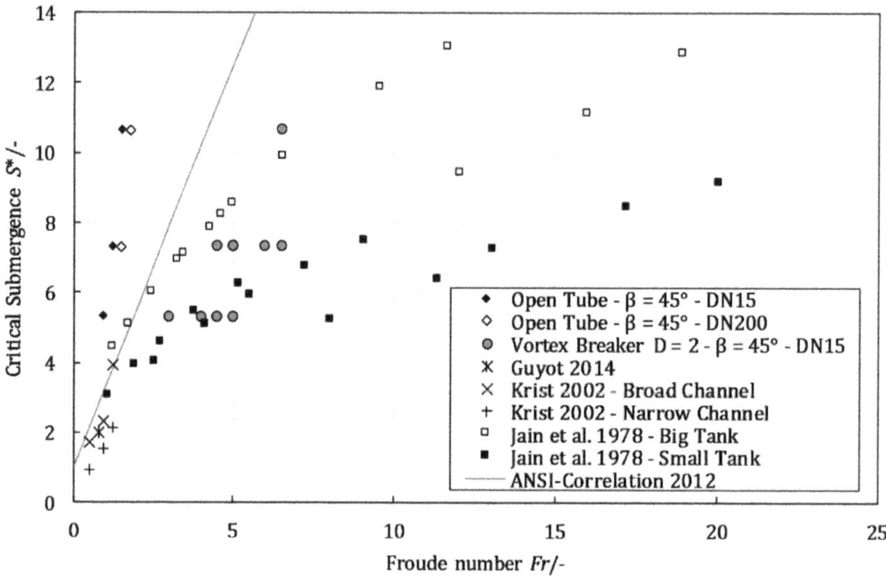

Figure 4-13: Comparison of experimental results with literature data. To enable the comparison of varying sizes, the critical submergence S^* is used plotted over the Froude number. The red line indicates the critical submergence defined by the ANSI-correlation, see eq. 2-46.

In Figure 4-13, the experimental results for the critical Froude number can be seen compared to literature data from *Guyot* [Guy14], *Jain et al.* [Jai788] and *Krist* [Kri02] as well as the ANSI correlation. Regarding the ANSI correlation, it is visible that it underestimates the critical submergence for almost all literature values, while the critical submergence for the experimental results in this study without vortex breaker are overestimated. All literature data are for low circulation systems, while the experimental values of this thesis included in the figure are for the highest measured circulations. When using vortex breaker, the Froude number corresponding to the critical submergence increases by a large margin, although the circulation in the system remains high. Note that only the critical values for the 2*D* vortex breaker of the DN15 laboratory setup could be compared, as the critical Froude numbers for the wide vortex breakers (4*D*) and the vortex breakers in the DN200 pilot plant lie outside the measurement range of the experimental setup.

As a result, the use of vortex breakers with cross or even better cross-baffle shape enable a more than 3.5 times higher volume flow rate compared to an open tube pump intake. The efficiency increase is even higher, but cannot be precisely determined, due to the limitations of the experimental setup.

4.2 Results of the Particle Image Velocimetry Measurements

The measurement of the horizontal velocity fields with PIV are conducted to serve two purposes. For one, the results can be used to visualize flow structures within and outside the vortex (core) region, which enables a qualitative comparison of the investigated pump intakes. Secondly, the results can be used to obtain quantitative values for the azimuthal velocity, circulation and the vortex core radius, by which the vortex development can be modelled with help of the theoretical vortex models described in chapter 2.2. Ultimately the goal is to use the quantitative and qualitative results for a common purpose: To increase the understanding of the underlying physical phenomena behind the vortex development and help to increase the accuracy of vortex predictions to improve the safety of operations.

For the qualitative description, Figure 4-14 shows exemplarily two-time averaged velocity fields within the vortex core region, recorded in the DN15 laboratory plant. Both are carried out for the same Froude number of Fr = 0.5 and the same measurement height. The difference between the recordings is that in the first case, the inflow angle is β = 45° and in the second case only β = 15°. Evidently, the axial velocity and therefore the vortex is much stronger in the first case, as the angular momentum induced into the system is much higher.

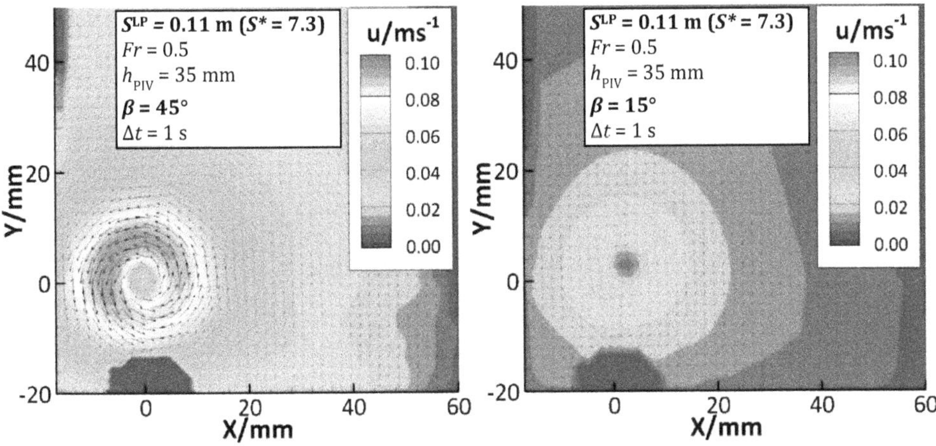

Figure 4-14: Comparison between two velocity fields in the DN15 laboratory plant for the same Froude number (Fr = 0.5) but different vertical inflow angles (left: β = 45° and right: β = 15°).

Since the velocities are subject to strong local and temporal fluctuations during the measurements, due to vortex movement and flow inhomgenities, the quantitative evaluation is carried out for each single image pair and not for the averaged files. Due to the stochastic movement of the vortex center, which can be seen in Figure 4-15, the origin of the measurement grid has to be adapted for each recorded frame. As the vortex stability is lower for smaller inflow angles, the vortex wanderings are strongest for recordings carried out for an inflow angle of β = 15°. The vortex center is most frequently found directly above the pump intake. The maximum deviation from this position grows with an increasing submergence depth and is also depending on the distance between the measurement plane and the pump intake. Although a larger Froude number is generally favorable for the vortex stability, turbulent substructures increase the instability due to the design of the inflow pipes, which lead to a deflection of the instreaming liquid from the wall and bottom of the vessel. These turbulent substructures can be visualized by numerical simulations conducted by *D. Bezecny* in the scope of the SAVE project [Sze17], see Figure 4-16.

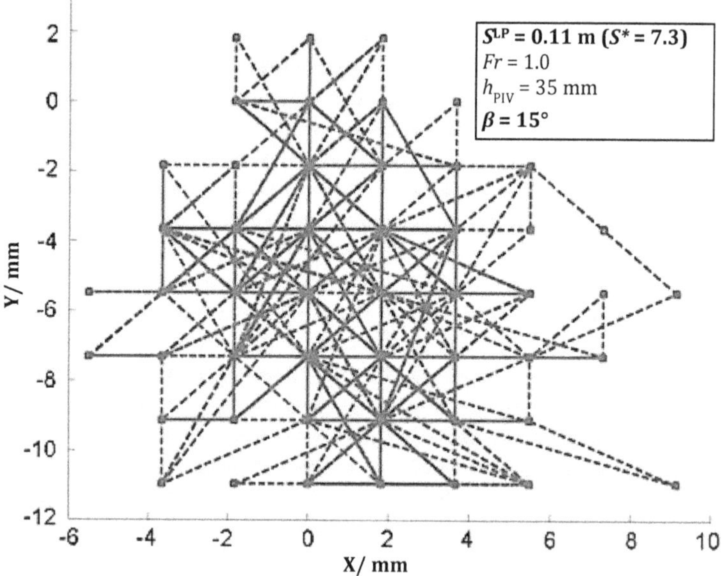

Figure 4-15: Stochastic movement of the vortex core in the DN15 laboratory plant for an inflow angle of β = 15° and Fr = 1.0. Each dot represents the position of the vortex center in a single image pair, while the lines represent the movement between two pairs. The grid like pattern is caused by the resolution of the Particle Image Velocimetry measurement.

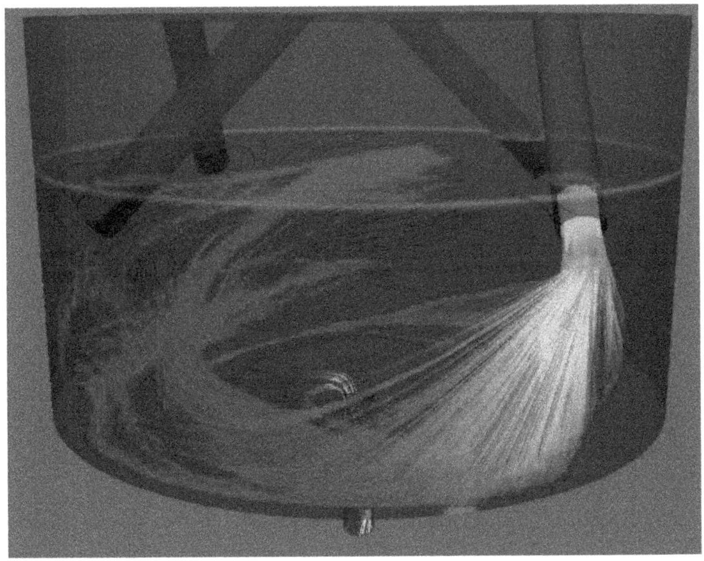

Figure 4-16: Numerical simulation of streamlines (blue) within the DN15 laboratory plant for $Fr = 1.2$ and an inflow angle of $\beta = 15°$, showing the turbulent substructures in the experimental setup due to the inflow hitting the wall and bottom of the cylindrical vessel. Simulations conducted by *D. Bezecny* [Sze17].

The azimuthal velocity profiles for the DN15 laboratory plant experiments are plotted in Figure 4-17 against the radius: In Figure 4-17a for a constant inflow angle of $\beta = 45°$ and varying Froude numbers and measurement heights, in Figure 4-17b for a Froude number of $Fr = 1.0$ and varying inflow angles. While the results are in good agreement with the PIV measurements conducted by *Caruso et al* [Car16] and velocity, maximum azimuthal velocity as well as circulation increase with increasing Froude number, the recorded velocities are generally very low and the radial value of the velocity maximum is too large, when compared to the dye-core experiments and the Laser Doppler Velocimetry measurements conducted with help of the *ILA R&D GmbH*. See Figure 4-18 for a comparison of the LDV and PIV measurements for $Fr = 0.5$. The main problems of the PIV measurement are a coarse particle seeding, a tendency of the particles to accumulate in the vortex center and the gas-core formation, which interferes with the vortex core and leads to reflections on the recordings. Due to the seeding problems, a wide grid must be taken to be able to evaluate the recordings, leading in turn to a strong averaging of the velocities within one grid cell an unfavorable resolution of the vortex core region. Indicated by the dye-core experiments, the vortex core is estimated to have a radius of ~1 – 2 mm, as can be seen in Figure 4-10a. The azimuthal velocity profile on the other side show a vortex core radius (the radius of the maximum velocity) of ~7.3 mm, which is the radius of the pump intake.

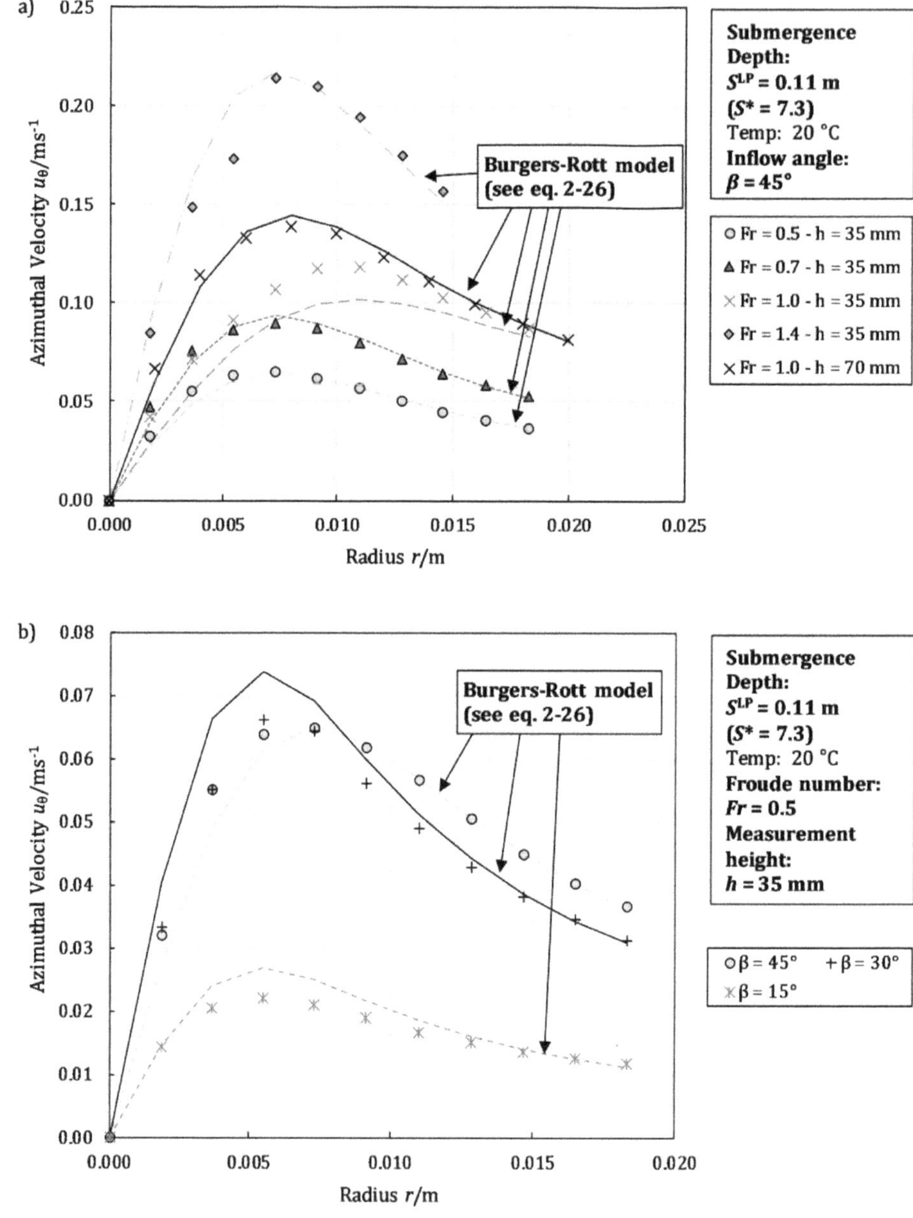

Figure 4-17: Azimuthal velocity profiles along the radius for PIV measurements conducted in the DN15 laboratory plant. a) For a constant inflow angle of β = 45° and b) for a constant Froude number of Fr = 0.5. Additionally, plotted are the velocity profiles calculated with the Burger-Rott model, based on the measured circulations and vortex core radii.

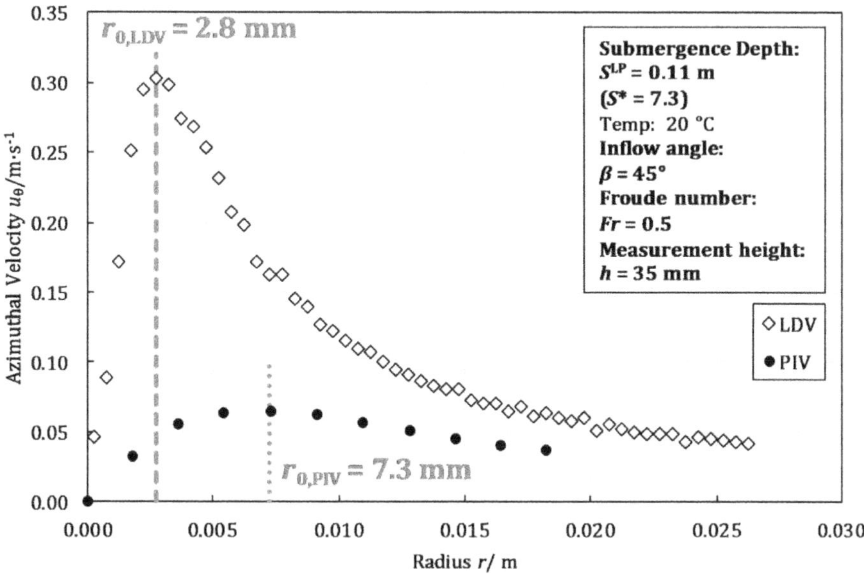

Figure 4-18: Comparison of LDV and PIV measurements conducted for the same experimental conditions in the DN15 laboratory plant.

From the PIV measurements in the DN200 pilot plant also a mean azimuthal velocity profile is derived and compared to numerical simulations (CFD) conducted by the *TÜV Nord Systec GmbH & Co. KG*, using *Ansys CFX* [Blö17]. Both, experimental and CFD results can be seen in Figure 4-19 plotted along the radius. Comparing the plots shows that the experimentally and numerically derived azimuthal velocities are in good agreement. Although the mean azimuthal velocities of the experiments in the vortex core region are in general a bit lower than the numerical ones, they are almost all within close proximity. For most measurement points, the numerical results of the nearest height are within the standard deviation of the experimental results. The experimentally determined velocities in the bulk phase are also a bit lower than the numerically determined velocities or the same Froude number, but again within the same range. Using eq. 2-7 and eq. 2-20 for the calculation of the bulk circulation and the downward acceleration results in $\Gamma_\infty^{\mathrm{exp}} = 0.3\ \mathrm{m^2s^{-1}}$ and $a^{\mathrm{exp}} = 2.7 \cdot 10^{-3}\ \mathrm{s^{-1}}$ for the experiments in the vortex core region and $\Gamma_\infty^{\mathrm{CFD}} = 0.4\ \mathrm{m^2s^{-1}}$ and $a^{\mathrm{CFD}} = 2.8 \cdot 10^{-3}\ \mathrm{s^{-1}}$ for the numerical simulation. From bulk circulation and axial acceleration, the theoretical gas-core length can be calculated for the experimental and numerical data, using eq. 2-28. Due to the difference in the values of the bulk circulation between experiments and simulations, the resulting theoretical gas-core lengths, calculated with the advanced Burgers-Rott model by *Ito et al.*, are $L^{BR,exp} = 0.11$ m and $L^{BR,CFD} = 0.20$ m. The difference in the circulation values is due to the general lower velocities measured in the experiment, resulting in an overall lower circulation, see Figure 4-20. While the PIV measurements lead to an underestimation of the gas-core lengths, the results of the numerical simulation are in very good agreement with the average measured gas-core length of $\bar{L}^{exp} = 0.19$ m or $\bar{L}^{*,exp} = 0.13$, proving that the Burgers-Rott model is applicable for this setup and can be used to predict the occurring gas-core lengths. Further prove is provided by numerical simulations, conducted by the *TÜV NORD EnSys GmbH & Co. KG* for $Fr = 1.0$ and $S = 1.46\ m$ [Blö17], by comparing the calculated surface curvature with the experimentally measured gas-core lengths, see Figure 4-21.

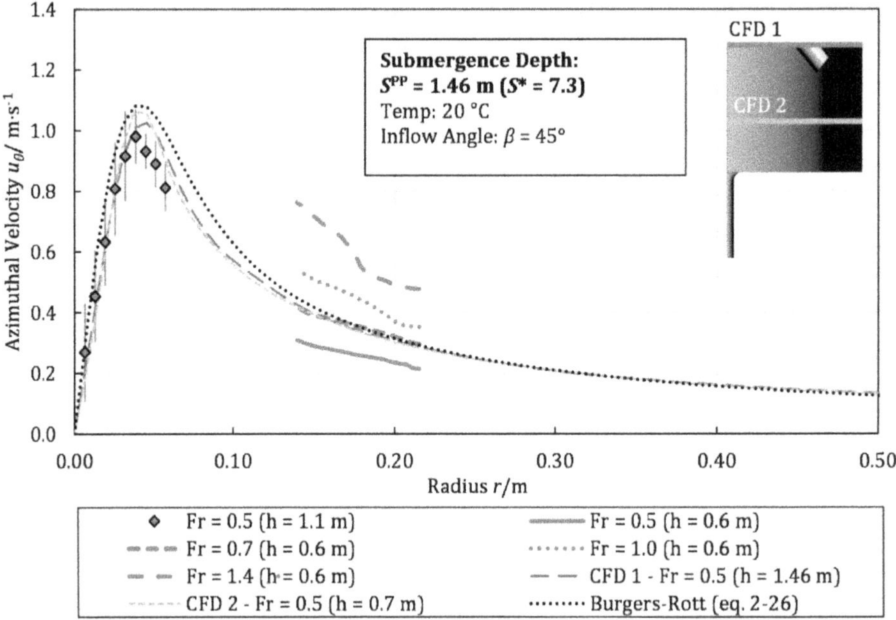

Figure 4-19: Comparison of PIV measurements and CFD simulations. Shown is the azimuthal velocity u_θ plotted over the radius r for an inflow angle $\beta = 45$ ° in the DN200 pilot plant. Red are the experimentally derived velocities, while blue and yellow are the velocities of the CFD simulations.

Overall, the experiments prove to be challenging due to the large-scale measurement setup. The largest issue is the loss of light, stemming from the long distances between the light source, measurement area and camera respectively and the limited field of view, making it necessary to conduct multiple measurements within the vortex core region and the bulk phase, to obtain the bulk circulation and vortex core radius. Additional problems are the same as in the DN15 laboratory setup: Seeding particles accumulate in the center of the vortex core, which leads to uneven particle distributions and an unresolvable vortex center, a coarse grid is needed for the evaluation due to uneven particle distributions and the recordings have to be descewed prior to evaluation. Ultimately, experiments could only be conducted successfully for $Fr = 0.5$, since the forming gas-core is superseding the vortex core for larger Froude numbers. Due to reflections at the forming gas-core, PIV measurements in the vicinity of the vortex core are rendered impossible.

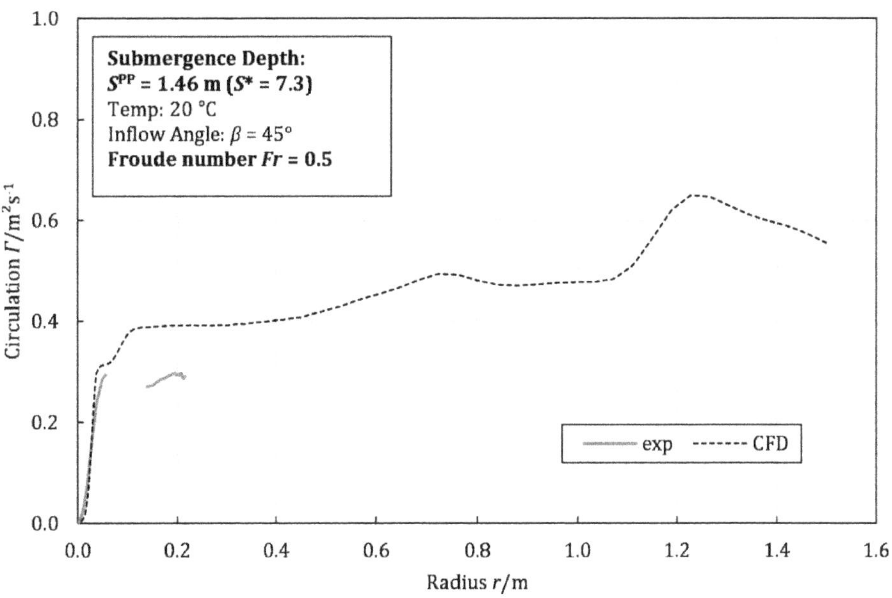

Figure 4-20: Experimentally and numerically derived circulation in the DN200 plant plotted over the radius for $Fr = 0.5$.

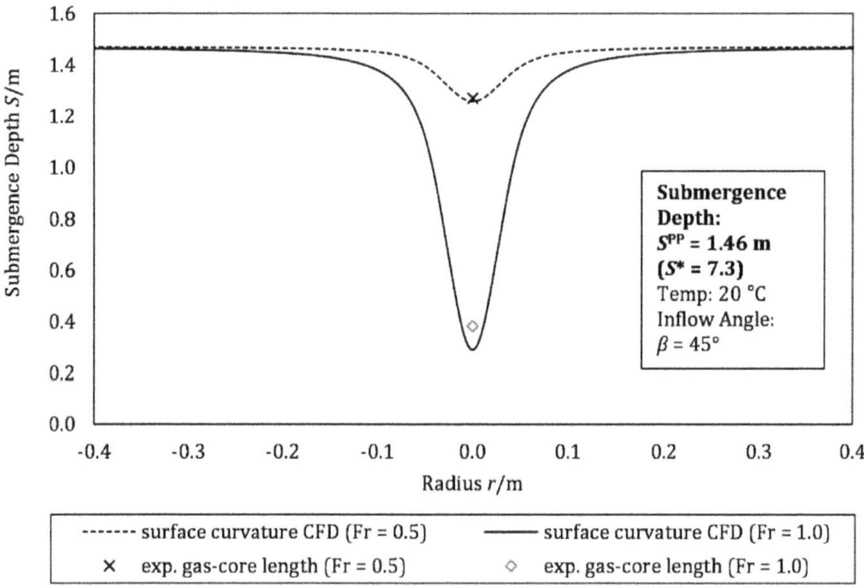

Figure 4-21: Comparison of surface curvatures and gas-core lengths in the DN200 pilot plant. Surface curvatures are derived from CFD results, gas-core lengths are determined experimentally.

4.3 Modelling

The results of the experiments conducted in the scope of this thesis and the research project SAVE, as well as the work of other authors, indicates clearly that due to the complexity of the physics involved in the vortex formation, no simple, universal correlation or model can be found, which would cover all the possible scenarios.
As shown in chapter 4.1.5, the ANSI correlation is too simplified, to be applicable for this experimental setup, yet other correlations (*Jain et al.* and *Knauss*) could not be applied, due to the unknown value of the circulation needed for the correlations.

The Burgers-Rott model on the other hand can be used successfully to predict the gas-core length within reasonable accuracy, as has been shown in the previous chapter. While the measurement of the circulation and the vortex core radius is possible, it is also very time consuming and challenging even in an experimental setup. Especially the vortex core radius is very hard to measure accurately, especially in the laboratory setup, due to its small value compared to the vessel's radius. A possible solution for this problem, besides numerical simulations which must be validated, is the combination of PIV with the dye experiments. Deriving the bulk circulation from the coarse gritted PIV measurements and the vortex core radius from the dye experiments.

For industrial applications, where no dye experiments can be conducted, the vortex core radius can also be estimated from the radius of the intake pipe. The experiments conducted in this thesis show that the vortex core radius and the intake pipe radius have the following correlation

$$r_0 = 0.1 \cdot D_{\text{Intake}} = 0.2 \cdot r_{\text{Intake}} \quad (4\text{-}6).$$

In other words, the vortex core radius is roughly 1/5th of the intake pipes radius wide. Inserting eq. 4-6 into eq. 2-29 leads to a new, simplified correlation for the gas-core length prediction

$$L = \frac{ln2}{g}\left(\frac{\Gamma_\infty}{0.4 \cdot \pi \cdot r_{\text{Intake}}}\right)^2 \approx 0.18 \cdot \left(\frac{\Gamma_\infty}{D}\right)^2 \quad (4\text{-}7),$$

which can also be rewritten in dimensionless form

$$L^* = \frac{ln2}{S \cdot g}\left(\frac{\Gamma_\infty}{0.4 \cdot \pi \cdot r_{\text{Intake}}}\right)^2 \approx \frac{0.18}{S} \cdot \left(\frac{\Gamma_\infty}{D}\right)^2 \quad (4\text{-}8),$$

This simplification, yet to be validated for other pump intake shapes and sizes, has the tremendous advantage that only the circulation in the bulk phase must be determined.

This can either be done by rather simple experiments, compared to the efforts needed to resolve the vortex core radius, or by numerical simulations. In Figure 4-22 the theoretical gas-core length predicted with eq. 4-8 are plotted against the measured gas-core lengths. It is visible that the gas-core length values are in good agreement with each other, although the correlation tends to overestimate smaller gas-core lengths.

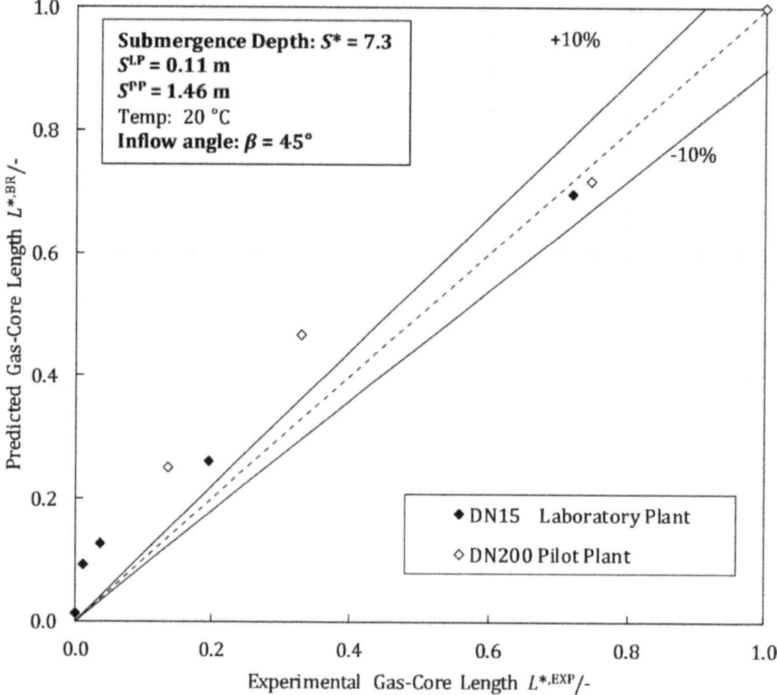

Figure 4-22: Validation of the simplified gas-core lengths prediction with the Burger-Rott model (eq. 4-8). In black for the DN15 laboratory plant, in white for the DN200 pilot plant. All experimental data is for experiments at $S^* = 7.3$, an inflow angle of $\beta = 45°$ and the Open Tube intake shape without vortex suppressors.

5 Conclusions

The results of the work conducted in this thesis, as well as the proposed model are a contribution to increase the predictability of vortex phenomena in pump intakes and to thereby decrease the threat of gas-entrainment into the pump systems. The experimental setup is chosen to generate a worst-case scenario with high circulations and an early onset of vortex formation, with strong and stable gas-cores.

The thesis can be separated into two main parts. In the first part, the vortex prevention by use of intake shapes and vortex breaker is investigated. Common vortex breaker types are investigated and evaluated regarding their efficiency in the prevention of gas-entrainment. Experiments conducted with different vortex suppressors show that the efficiency of cross and cross-plated/baffle-plated vortex breakers in preventing gas-entrainment into the pump system even is depending to a large degree on their diameter. Wider vortex breakers drastically hinder the gas-core formation to the point that no critical Froude number can be reached within the observable measurement range. The most efficient vortex-breaker, a wide cross-plate with a diameter of four times the intake diameter ($4D$), enables a more than five times higher volume flow rate to be pumped. Although the efficiency increase is most likely much higher, since the no gas-entrainment is recordable, when the limits of the experimental setup are reached.

In the second part of the thesis, the focus is set on the vortex formation under high circulations. The measurement results for an open, unmodified pump intake show a sigmoid-shaped development of the gas-core, with a strong dependency on the Froude number, which can be formed into a correlation to predict the gas-core length, depending mostly on the Froude number and the submergence depth.

Additionally, Particle Image Velocimetry measurements to determine the velocity field inside the vortex core region are conducted. For the measurements in the DN200 pilot plant scale, a novel High-Power LED system is implemented, improving the safety of the conducted experiments. The PIV results, in combination with dye experiments, are used to model an equation for the prediction of gas-core lengths, based on the theoretical vortex model of Burgers-Rott with the improvement made by Ito et al. and to validate numerical simulations.

Bibliography

[Anw78] Anwar; Weller; Amphlett: Similarity of Free-Vortex at Horizontal Intake, J IAHR, No. 2, 1978

[Anw80] Anwar; Amphlett: Vortices at Vertically Inverted Intake, J. IAHR, No. 2, 1980

[Auc09] Auckland, A.; Nistor, I.; Townsend, R.: Intake shape and boundary-related considerations in the operation and design of storm-sewer drop structures, Can. J. Civ. Eng., 2009, doi:10.1139/L09-114

[Blö17] Blömeling, F., and Lawall, R., (2017). "Schlussbericht Verbundprojekt SAVE: Sicherheitsrelevante Analyse des Verhaltens von Armaturen, Kreiselpumpen und Einlaufgeometrien unter Berücksichtigung störfallbedingter Belastungen Teilprojekt C: Analytische Untersuchungen der Wirbelbildung in Pumpeneinläufen [Final Report Project SAVE: Safety-relevant Analysis of the Performance of Centrifugal Pumps, Valves and Inlet Geometries, including stress-related Events. Subproject C: Analytical Investigation of Vortex Development in Pump Intakes], TÜV NORD EnSys GmbH & Co. KG, Hamburg, Germany (in German)

[Bor10] Borgei, S. M., and Kabiri-Samani, A. R.: Effect of Anti-Vortex Plates on Critical Sub-mergence at a Vertical Intake, Scientia Iranica, Transaction A: Civil Engineering, 17(2), 89-95, 2010

[Bur48] Burgers, J. M.: A Mathematical Model Illustrating the Theory of Turbulence, Adv. Appl. Mech., Vol. 1, 1948

[Car16] Caruso, G.; Cristofano, L.; Nobili, M.; Romano, G.: Experimental Investigation on Free Surface Vortices Driven by Tangential Inlets, Int. J. Heat Mass Transf. Vol 34, No. 4, 2016

[Dag74] Daggett, L. L.; Keulegan, G. H.: Similitude Conditions in Free-Surface Vortex Formations, J. Hydr. Eng Div, No. 100, 1974

[Dhi80] Dhillon: Vortex Formation at Pipe Intakes and its Prediction, CBIP New Dehli, Status Report No. 6, 1980

[Ekh12] Ekholm, M.: Screen basket vortex breaker for vessel, Patent-№ US8439071B2, 2012, https://patents.google.com/patent/US8439071B2/en

[Gra66] Granger, R.: Steady three-dimensional vortex flow, J. Fluid. Mech. Vol. 25, pt. 3, pp. 557-576, 1966

[Gül13] Gülich, J. F.: Kreiselpumpen. 4. aktual. u. erw. Aufl., Springer-Verlag Berlin Heidelberg, 2013

[Gul83] Gulliver, J. S.; Rindels, A. J.: An Experimental Study of Critical Submergence to Avoid Free-surface Vortices at Vertical Intakes, Report for the Legislative Commission on Minnesota Resources, Nr. 224, 1983

[Guy14] Guyot, G.; Maaloul, H.; Archer, A.: A Vortex Modeling with 3D CFD, Advances in Hydroinformatics, Springer Hydrology, 2014, DOI: 10.1007/978-981-4451-42-0_35

[Hec78] Hecker, G. E.; Durgin, W. W.: The Modelling of Vortices at intake Structures, IAHR-ASME-ASCE Joint Symposium on Design and Operation of Fluid Machinery, CSU Fort Collins, USA, 1978

[Hyd12] American National Standard for Pump Intake Design – Rotodynamic Pumps, Hydraulic Institute Standards, 2012

[Ito10] Ito, K.; Sakai, T.; Eguchi, Y.; Monji, H.; Ohshima, H.; Uchibori, A. Xu, Y.: Improvement of Gas Entrainment Prediction Method, Introduction of Surface Tension Effect, J. Nucl. Sci. Technol., Vol. 47(9), 2010

[Jai78] Jain, A.K.; Ranga Raju, K. G.; Garde, R. J.: Vortex Formation at Vertical Pipe Intakes, ASCE, Hydraulics Division, HY10, 1978

[Kir12] Kirst, K.: Experimentelle und numerische Untersuchungen von Zulaufbedingungen vertikaler Pumpsysteme (dissertation), Shaker Verlag, Aachen, 2012 (in German)

[Kis90] Kister, H.: Distillation Operations, McGraw-Hill, New York, USA, p. 91, 1990, ISBN 9780070349100

[Kel14] Keller, J.; Möller, G.; Boes, R. M.: PIV Measurements of Air-core Intake Vortices, Flow Meas. Instrum., Vol. 40, 2014

[Kna83] Knauss, J.: DVWK Schriften: Wirbelbildung an Einlaufbauwerken – Luft- und Dralleintrag (German), Kommissionsvertrieb Verlag Paul Parey, Hamburg und Berlin, Germany, 1983

[Lis00] List, M.; Kinzy, J.: Anti-vortex baffle assembly with filter for a tank, Patent-№ US6014987A, 2009, https://patents.google.com/patent/US6014987A/en

[Lub67] Lubin, B.; Springer, G. S.: The formation of a dip on the surface of a liquid draining from a tank, J. Fluid Mech., Vol. 29, 1967

[Lud99] Ludwig, E.; Applied Process Design for Chemical and Petrochemical Plants: Volume 1, 3rd Edition, Butterworth-Heinemann, Woburn, USA, p.189, 1999, ISBN 0884150259

[Mah10] Mahyari, M.; Karimi, H.; Naseh, H.; Mirshams, M.: Numerical and experimental investigation of vortex breaker effectiveness on the improvement in launch vehicle ballistic parameters, J Mech Sci Technol, Vol. 24 (10), 2010, DOI 10.1007/s12206-010-0618-7

[Man95] Manning, F.; Thompson, R.: Oilfield Processing Volume Two: Crude Oil, PenWell Publishing, Tulsa, USA, p. 88, 1995, ISBN 0878143548

[Mor09] Morrissey, M.; Kelly, J.; Wong, D.; Mello, S.; Weissenbach, J.-L.: Container having vortex breaker and system, Patent-№ US9290729B2, 2009, https://patents.google.com/patent/US9290729B2/en

[Nog03] Noguchi, T; Yukimoto, S.; Kimura, R.; Niino, H.: Structure and instability of a sink vortex, Proceedings of PSFVIP-4, Chamonix, France, June 3-5 2003

[Odg86] Odgaard, A. J.: Free-surface Air Core Vortex, J. Hydraul. Eng., Vol. 112, 1986

[Oga00] Ogawa, A.; Ugai, T.: Collection Efficiency of Cylindrical and Elliptic Cyclone Dust Collectors with Vortex Breaker, J Therm Sci, Vol. 9 (3), 2000

[Pad82] Padamanabhan, M.: Evaluation of Vortex Breakers, Hydraulic Performance of Single Outlet Sumps, and Sensitivity of Miscellaneous Sump Parameters, Alden Research Laboratory Report, 49A-82/M398F, Holden, MA, 1982

[Pan13] Pandazis, P. Blömeling, F.: Final report: Joint research project GUMP: Generic numerical determination of the critical submergence at pump intakes to avoid gas entrainment, WP3: Investigation of influencing parameters on the formation of surface vortices, Technical Report. Reactor Safety Research – Project No.: 1501410, 2013 (in German)

[Raf07] Raffel, M.; Willert, C.; Wereley, S.; Kompenhans, J.: Particle Image Velocimetry – A Practical Guide, Springer Publishing, Berlin, Germany, 2007, 978-3-540-72307-3

[Rah87] Rahlwes, W.: Vortex breaker for use in a liquid-liquid separator or the like, Patent-№ US4696741A, 1987, https://patents.google.com/patent/US4696741A/en

[Ran58] Rankine, W. J. M.: A Manual of Applied Mechanics, Richard Griffin and Company London Glasgow, 1858

[Rei82] Reimann, J.; Khan, M.: Flow through a Small Pipe at the Bottom of a Large Pipe with Stratified Flow, Annual Meeting of the European Two-Phase-Flow Group, Paris-La defense, 1982

[Rot58] Rott, N.: On the Viscous Core of a Line Vortex, Zeitschrift für angewandte Mathematik und Physik ZAMP, Vol. 9(5-6), 1958

[Stri92] Strickland, J.; Steinberg, R.: Vortex breaker for horizontal liquid draw off tray sump, Patent-№ US5096578A, 1992, https://patents.google.com/patent/US5096578A/en

[Sze15] Szeliga, N.; Richter, S.; Bezecny, D.; Schlüter, M.: Determination of tangential velocities in free surface vortices by high-speed PIV measurements, in proceedings of the "23. Fachtagung Lasermethoden in der Strömungsmesstechnik in Dresden", Gala e.V., Karlsruhe, Germany, pp. 43-1 – 43-10, ISBN: 978-3-9816764-1-9, 2015

[Sze16] Szeliga, N.; Richter, S.; Bezecny, D.; Hoffmann, M.; Schlüter, M.: Determination of the Influence of Tangential Momentum on Air-Core Vortex Formation at Pump Intakes by Means of Particle Image Velocimetry, in proceedings of the "24. Fachtagung Lasermethoden in der Strömungsmesstechnik in Cottbus", Gala e.V., Karlsruhe, Germany, pp. 30-1 – 30-9, ISBN: 978-3-9816764-2-6, 2016

[Sze16b] Szeliga, N.; Richter, S.; Bezecny, D.; Hoffmann, M.; Schlüter, M., Schäfer, T.; Hampel, U.; Blömeling, F.; Lawall, R.; Hamberger, M.; Walberer, A.: BMBF-Projekt SAVE: Sicherheitsrelevante Analyse des Verhaltens von Armaturen, Kreiselpumpen und Einlaufgeometrien unter Berücksichtigung störfallbedingter Belastungen, in proceedings of Kraftwerkstechnik 2016, SAXONIA Standortetwicklungs- und –verwaltungsgesellschaft mbH, Freiberg, Germany, pp. 745 – 758, ISBN: 978-3-934409-73-6, 2016 (in German)

[Sze17] Szeliga, N.; Bezecny, D.; Richter, S.; Schlüter, M.; Final report: Joint research project SAVE: Safety-relevant analysis of the performance of centrifugal pumps, valves and inlet geometries, including stress-related events, subproject A, Hamburg, Germany, https://doi.org/10.2314/GBV:1011865149, 2017 (in German)

[Sze19] Szeliga, N.; Helmrich von Elgott, L.; Bezecny, D.; Richter, S.; Hoffmann, M.; Schlüter, M.: Large Scale Experiments on the Formation of Surface Vortices with and without Vortex Suppression, Chemie Ingenieur Technik, 2019, https://doi.org/10.1002/cite.201900091

[Tri10] Trivellato, T.: Anti-vortex devices: Laser measurements of the flow and functioning, Opt. Lasers Eng., 48, 589–599, 2010

[VDI10] VDI Heat Atlas, Springer Link, 2010. ISBN: 978-3-540-79999-3

[Wag05] Wagner, W.: Kreiselpumpen und Kreiselpumpenanlagen, Vogel Publishing, 2005

[Wu06] Wu, J.Z.; Ma, H.Y.; Zhou, M.-D.: Vorticity and Vortex Dynamics, Springer Publishing, 2006

[Yan14] Yang, J.; Liu, T.; A. Bottacin-Busolin, C. Lin, Effects of intake-entrance profiles on free-surface vortices, J. Hydraul. Res., Vol. 52(4), 2014

[Zlo06] Zlokarnik, M.: Scale-up in Chemical Engineering. Weinheim, Germany, Wiley-VCH Verlag GmbH, 2012

Publications

Paper:

N. Szeliga et al., *BMBF-Verbundprojekt SAVE: Sicherheitsrelevante Analyse des Verhaltens von Armaturen, Kreiselpumpen und Einlaufgeometrien unter Berücksichtigung störfallbedingter Belastungen*, 48. Kraftwerkstechnisches Kolloquium, 18.-19.10.2016, Dresden, Deutschland. Im Tagungsband veröffentlicht, ISBN: 978-3-934409-69-9

N. Szeliga, S. Richter, D. Bezecny, M. Hoffmann, M. Schlüter, *Determination of the Influence of Tangential Momentum on Air-Core Vortex Formation at Pump Intakes by Means of Particle Image Velocimetry*, 24. Fachtagung "Experimentelle Strömungsmechanik" GALA e.V., 6.-8.09.2016, Cottbus, Deutschland. Im Tagungsband veröffentlicht, ISBN: 978-3-9816764-2-6

N. Szeliga, S. Richter, D. Bezecny, M. Schlüter, *Determination of tangential velocity in free surface vortices by high-speed PIV measurements*, 23. Fachtagung "Lasermethoden in der Strömungsmechanik", GALA e.V., 8.-10.09.2015, Dresden, Deutschland. Im Tagungsband veröffentlicht, ISBN: 978-3-9816764-1-9

Szeliga, N.; Helmrich von Elgott, L.; Bezecny, D.; Richter, S.; Hoffmann, M.; Schlüter, M.: *Large Scale Experiments on the Formation of Surface Vortices with and without Vortex Suppression*, Chemie Ingenieur Technik, 2019, https://doi.org/10.1002/cite.201900091

Oral:

N. Szeliga, D. Bezecny, S. Richter, M. Hoffmann, M. Schlüter, *Reliable prediction of air-entraining vortices at pump intakes to prevent the failure of cooling systems (BMBF-Project SAVE).* CHISA Conference 2018 in Prague, CZ

N. Szeliga, D. Bezecny, S. Richter, M. Hoffmann, M. Schlüter, *Experimentelle und numerische Untersuchungen zur Vermeidung von Gasmitriss in Notkühlsystemen: Das BMBF-Verbundprojekt SAVE,* ACHEMA 2018 Kongress 11.-15.06.2018, Frankfurt, Deutschland

N. Szeliga, S. Richter, D. Bezecny, M. Hoffmann, M. Schlüter, *Influence of angular momentum on the formation of air-core intake vortices in cylindrical water tanks,* FSP-Workshop "Klimaschonende Energie- und Umwelttechnik", 30.11.2016, Hamburg, Deutschland

N. Szeliga et al., *BMBF-Verbundprojekt SAVE: Sicherheitsrelevante Analyse des Verhaltens von Armaturen, Kreiselpumpen und Einlaufgeometrien unter Berücksichtigung störfallbedingter Belastungen*, 48. Kraftwerkstechnisches Kolloquium, 18.-19.10.2016, Dresden, Deutschland

N. Szeliga, S. Richter, D. Bezecny, M. Hoffmann, M. Schlüter, *Determination of the Influence of Tangential Momentum on Air-Core Vortex Formation at Pump Intakes by Means of Particle Image Velocimetry*, 24. Fachtagung "Experimentelle Strömungsmechanik" GALA e.V., 6.-8.09.2016, Cottbus, Deutschland

N. Szeliga, S. Richter, D. Bezecny, M. Hoffmann, M. Schlüter, *Influence of angular momentum on the formation of air-core intake vortices in cylindrical water tanks*, ICMF 2016 International Conference on Multiphase Flow, 22.-27.05.2016, Florenz, Italien

N. Szeliga, S. Richter, D. Bezecny, M. Schlüter, *Determination of tangential velocity in free surface vortices by high-speed PIV measurements*, 23. Fachtagung "Lasermethoden in der Strömungsmechanik", GALA e.V., 8.-10.09.2015, Dresden, Deutschland

N. Szeliga, S. Richter, D. Bezecny, M. Hoffmann, M. Schlüter, *Sicherheitsrelevante Analyse des Verhaltens von Armaturen, Kreiselpumpen und Einlaufgeometrien unter Berücksichtigung störfallbedingter Belastungen*, 2. Projektstatusgespräch zur BMBF-geförderten Nuklearen Sicherheitsforschung, 25.-26.03.2015, Dresden, Deutschland

Poster:

N. Szeliga, S. Richter, D. Bezecny, M. Hoffmann, M. Schlüter, *Reliable prediction of air-entraining vortices at pump intakes to prevent the failure of cooling systems (BMBF-Project SAVE)*, FSP-Workshop "Klimaschonende Energie- und Umwelttechnik", 2017, Hamburg, Deutschland

Supervised Student Theses

Name	Year	Title
Hans-Christian Lewitz Bachelor thesis	2015	Experimentelle Ermittlung der Luftkernlängen von Hohlwirbeln unter Variation der Ansauggeometrie und Froude-Zahl
Christian Born Bachelor thesis	2015	Experimentelle Analyse von Scale-up Regeln für Pumpenansaugbecken mit Hilfe eines MATLAB-Bildanalyse Tools
Nathali del Carmen Schmid Bachelor thesis	2015	Experimentelle Untersuchung der Tangentialgeschwindigkeiten von luftziehenden Hohlwirbeln in Pumpenansaugbecken verschiedener Größe
Cihan Çantay Bachelor thesis.	2015	Experimentelle Untersuchung von wirbelbrechenden Maßnahmen im Ansaugbecken großtechnischer Kühlkreisläufe
Laura Westermann Master thesis	2016	Experimentelle Untersuchung des Einflusses tangentialer Impulseinträge auf die Ausbildung luftziehender Hohlwirbel im Labor- und Industriemaßstab
Felix Kexel Bachelor thesis	2016	Experimentelle Untersuchung der Auswirkungen von luftziehenden Hohlwirbeln auf die Kreiselpumpe eines Notkühlsystems im Industriemaßstab
Lennard Ole Haskamp Bachelor thesis	2017	Scale-up of liquid-liquid reactors based on the droplet size distribution measurements under laboratory and industrial conditions
Julián A. Giraldo-Ospina Master thesis	2018	Experimentelle Analyse und energetische Bewertung einer Reifenvulkanisiermaschine
Nadja Kenndoff Master thesis	2019	Characterization and Modelling of the Droplet Size Distribution in a Stirred Tank Reactor Based on the Energy Dissipation Density and Comparison with a Free Jet
Lando Helmrich von Elgott Bachelor thesis	2019	Vermeidung von Gaseintrag in Pumpensystemen - Einfluss der Abflussgeometrie auf die Ausbildung von Hohlwirbeln und das Strömungsfeld einer Wirbelströmung
Felix Kexel Master thesis	2019	Design and Characterization of a 3D-Printed Jet Loop Reactor with Optimized Flow Conditions for Enhanced Mass Transfer
Mauricio Vaca Guerra Master thesis	2019	Characterization and Modeling of the Droplet Size Distribution within a liquid-liquid Jet stream of a Jet Loop Reactor.
Darwin Schnute Bachelor thesis	2019	Optimierungspotenzial von Treibstrahl-Schlaufenreaktoren mittels 3D gedruckter Einbauten

Appendix

A-1: DN15 laboratory plant volume flow rates, velocities and dimensionless numbers.

					DN15 Laboratory Plant			
$Q/m^3 \cdot h^{-1}$	$\dot{M}/kg \cdot s^{-1}$	$Q/m^3 \cdot s^{-1}$	$Q/L \cdot min^{-1}$	$u/m \cdot s^{-1}$	Fr/-	Re_r/-	Re/-	We/-
0.02	6.65E-03	6.78E-06	0.41	0.04	**0.1**	62	574	0.55
0.05	1.33E-02	1.36E-05	0.81	0.08	**0.2**	123	1149	1.09
0.07	1.99E-02	2.03E-05	1.22	0.12	**0.3**	185	1723	1.64
0.10	2.66E-02	2.71E-05	1.63	0.15	**0.4**	246	2297	2.18
0.12	3.32E-02	3.39E-05	2.03	0.19	**0.5**	308	2871	2.73
0.15	3.99E-02	4.07E-05	2.44	0.23	**0.6**	369	3446	3.27
0.17	4.65E-02	4.75E-05	2.85	0.27	**0.7**	431	4020	3.82
0.20	5.32E-02	5.42E-05	3.25	0.31	**0.8**	492	4594	4.36
0.22	5.98E-02	6.10E-05	3.66	0.35	**0.9**	554	5168	4.91
0.24	6.65E-02	6.78E-05	4.07	0.38	**1.0**	615	5743	5.46
0.27	7.31E-02	7.46E-05	4.47	0.42	**1.1**	677	6317	6.00
0.29	7.98E-02	8.13E-05	4.88	0.46	**1.2**	738	6891	6.55
0.32	8.64E-02	8.81E-05	5.29	0.50	**1.3**	800	7465	7.09
0.34	9.31E-02	9.49E-05	5.69	0.54	**1.4**	861	8040	7.64
0.37	9.97E-02	1.02E-04	6.10	0.58	**1.5**	923	8614	8.18
0.39	1.06E-01	1.08E-04	6.51	0.61	**1.6**	984	9188	8.73
0.41	1.13E-01	1.15E-04	6.91	0.65	**1.7**	1046	9762	9.27
0.44	1.20E-01	1.22E-04	7.32	0.69	**1.8**	1107	10337	9.82
0.46	1.26E-01	1.29E-04	7.73	0.73	**1.9**	1169	10911	10.37
0.49	1.33E-01	1.36E-04	8.13	0.77	**2.0**	1230	11485	10.91
0.51	1.40E-01	1.42E-04	8.54	0.81	**2.1**	1292	12059	11.46
0.54	1.46E-01	1.49E-04	8.95	0.84	**2.2**	1353	12634	12.00
0.56	1.53E-01	1.56E-04	9.35	0.88	**2.3**	1415	13208	12.55
0.59	1.60E-01	1.63E-04	9.76	0.92	**2.4**	1476	13782	13.09
0.61	1.66E-01	1.69E-04	10.17	0.96	**2.5**	1538	14356	13.64
0.63	1.73E-01	1.76E-04	10.57	1.00	**2.6**	1599	14931	14.18
0.66	1.80E-01	1.83E-04	10.98	1.04	**2.7**	1661	15505	14.73
0.68	1.86E-01	1.90E-04	11.39	1.07	**2.8**	1722	16079	15.28
0.71	1.93E-01	1.97E-04	11.80	1.11	**2.9**	1784	16653	15.82
0.73	1.99E-01	2.03E-04	12.20	1.15	**3.0**	1845	17228	16.37
0.85	2.33E-01	2.37E-04	14.24	1.34	**3.5**	2153	20099	19.09
0.98	2.66E-01	2.71E-04	16.27	1.53	**4.0**	2460	22970	21.82
1.10	2.99E-01	3.05E-04	18.30	1.73	**4.5**	2768	25841	24.55
1.22	3.32E-01	3.39E-04	20.34	1.92	**5.0**	3075	28713	27.28
1.34	3.66E-01	3.73E-04	22.37	2.11	**5.5**	3383	31584	30.01
1.46	3.99E-01	4.07E-04	24.40	2.30	**6.0**	3690	34455	32.73
1.59	4.32E-01	4.41E-04	26.44	2.49	**6.5**	3998	37326	35.46

A-2: DN200 pilot plant volume flow rates, velocities and dimensionless numbers.

DN200 Pilot Plant							
$Q/m^3 \cdot h^{-1}$	$\dot{M}/kg \cdot s^{-1}$	$Q/m^3 \cdot s^{-1}$	$u/m \cdot s^{-1}$	Fr/-	Re_r/-	Re/-	We/-
16	4.32	4.40E-03	0.14	**0.1**	3002	27958	7.27
32	8.63	8.80E-03	0.28	**0.2**	6004	55917	14.55
48	12.95	1.32E-02	0.42	**0.3**	9006	83875	21.82
63	17.27	1.76E-02	0.56	**0.4**	12007	111833	29.10
79	21.58	2.20E-02	0.70	**0.5**	15009	139792	36.37
95	25.90	2.64E-02	0.84	**0.6**	18011	167750	43.64
111	30.22	3.08E-02	0.98	**0.7**	21013	195709	50.92
127	34.53	3.52E-02	1.12	**0.8**	24015	223667	58.19
143	38.85	3.96E-02	1.26	**0.9**	27017	251625	65.47
158	43.17	4.40E-02	1.40	**1.0**	30018	279584	72.74
174	47.49	4.84E-02	1.54	**1.1**	33020	307542	80.02
190	51.80	5.28E-02	1.68	**1.2**	36022	335500	87.29
206	56.12	5.72E-02	1.82	**1.3**	39024	363459	94.56
222	60.44	6.16E-02	1.96	**1.4**	42026	391417	101.84
238	64.75	6.60E-02	2.10	**1.5**	45028	419375	109.11
253	69.07	7.04E-02	2.24	**1.6**	48029	447334	116.39
269	73.39	7.48E-02	2.38	**1.7**	51031	475292	123.66
285	77.70	7.92E-02	2.52	**1.8**	54033	503251	130.93
301	82.02	8.36E-02	2.66	**1.9**	57035	531209	138.21
317	86.34	8.80E-02	2.80	**2.0**	60037	559167	145.48
333	90.65	9.24E-02	2.94	**2.1**	63039	587126	152.76
349	94.97	9.68E-02	3.08	**2.2**	66040	615084	160.03
364	99.29	1.01E-01	3.22	**2.3**	69042	643042	167.31
380	103.60	1.06E-01	3.36	**2.4**	72044	671001	174.58
396	107.92	1.10E-01	3.50	**2.5**	75046	698959	181.85
412	112.24	1.14E-01	3.64	**2.6**	78048	726917	189.13
428	116.56	1.19E-01	3.78	**2.7**	81050	754876	196.40
444	120.87	1.23E-01	3.92	**2.8**	84051	782834	203.68
459	125.19	1.28E-01	4.06	**2.9**	87053	810793	210.95
475	129.51	1.32E-01	4.20	**3.0**	90055	838751	218.22
507	138.14	1.41E-01	4.48	**3.2**	96059	894668	232.77
539	146.77	1.50E-01	4.76	**3.4**	102063	950584	247.32
570	155.41	1.58E-01	5.04	**3.6**	108066	1006501	261.87
602	164.04	1.67E-01	5.32	**3.8**	114070	1062418	276.42
634	172.67	1.76E-01	5.60	**4.0**	120074	1118335	290.97
713	194.26	1.98E-01	6.30	**4.5**	135083	1258126	327.34
792	215.84	2.20E-01	7.00	**5.0**	150092	1397918	363.71
871	237.43	2.42E-01	7.70	**5.5**	165101	1537710	400.08
951	259.01	2.64E-01	8.40	**6.0**	180110	1677502	436.45

MegaCPK Data Sheet

Data sheet

Customer item no.:
Order dated: 18/01/2013
Order no.: Anfrage TU HH
Quantity: 1

Number: ES 2149507
Item no.: 300
Date: 18/01/2013
Page: 1 / 6

Version no.: 2

CPKN-C1 300-500
Chemical pump to EN 22858/ISO 2858/ISO 5199

Operating data

Requested flow rate	1000.00 m³/h	Actual flow rate	1000.00 m³/h
Requested developed head	26.00 m	Actual developed head	26.00 m
Pumped medium	Water, boiler feed water Fully desalinated water treated to VdTÜV 1466 Not containing chemical and mechanical substances which affect the materials	Efficiency	84.7 %
		Power absorbed	83.46 kW
		Pump speed of rotation	889 rpm
		NPSH required	3.36 m
		Permissible operating pressure	16.00 bar.g
Fluid temperature	20.0 °C	Discharge press	2.54 bar.g
Fluid density	998 kg/m³		
Fluid viscosity	1.00 mm²/s		
Suction pressure max.	0.00 bar.g	Min. allow. mass flow rate for stable operation	82.38 kg/s
Mass flow rate	277.22 kg/s		
Max. power on curve	87.01 kW	Shutoff head	31.95 m
Min. allow. flow rate for stable operation	297.17 m³/h	Performance test	Yes

Design

Pump standard	ISO 5199	Type	H75N
Design	Baseplate mounted, long-coupled	Material code	AQ1EGG
Orientation	Horizontal	Sealing plan	E Single acting mechanical (external circulation)
Shaft execution	Dry	Minimum requirements for hot water quality: treatment acc. to VdTÜV regulation TCH 1466 With SiO2-content up to max. 10 mg/l. Conductivity max. 50 µS/cm for water with low salt content up to 160°C or conductivity max. 250 µS/cm up to 140°C for water with higher salt content Solids content up to max. 5 mg/l and no additives forming a greasy film on the seal faces.	
Suction nominal dia.	DN 350		
Suction nominal pressure	PN 16		
Suction position	axial		
Connection standard suction	EN 1092-1		
Discharge nominal dia.	DN 300		
Discharge nominal pressure	PN 16		
Discharge position	top (0°/360°)	Seal chamber design	Standard seal chamber
Connection standard discharge	EN 1092-1	Impeller diameter	504.0 mm
Surface type	Raised face (B / RF)	Direction of rotation from drive	Clockwise
Shaft seal	Single acting mechanical seal	Bearing bracket construction	Reinforced (heavy)
Manufacturer	Burgmann	Bearing bracket size	P08s
		Bearing seal	Standard labyrinth ring
		Bearing type	Anti-friction bearings
		Lubrication type	Oil
		Lubrication monitoring	Constant level oiler
		Color	Ultramarine blue (RAL 5002) KSB-blue

Data sheet

Customer item no.:
Order dated: 18/01/2013
Order no.: Anfrage TU HH
Quantity: 1

Number: ES 2149507
Item no.: 300
Date: 18/01/2013
Page: 2 / 6

Version no.: 2

CPKN-C1 300-500
Chemical pump to EN 22858/ISO 2858/ISO 5199

Driver, accessories

Coupling type	Metastream TSK	Motor speed	988 rpm
Nominal size	135	Frequency	50 Hz
Spacer length	200.0 mm	Operating voltage	400 V
Coupling guard type	Tread-proof (ZN3230)	Rated power P2	110.00 kW
Guard size	A2	Rated current	196.0 A
Guard material	Steel	Starting current ratio	7
Baseplate type	Welded steel according to KSB AN 845	Insulation class	F to IEC 34-1
		Motor enclosure	IP55
Baseplate size	17	Cos phi at 4/4 load	0.86
Baseplate fixing	Foundation bolts galvanized steel	Temperature sensor	3 PTC resistors
		Terminal box position	0°/360° (top) Viewed from the drive
Driver type	Electric motor	Motor winding	400 / 690 V
Model (make)	Siemens	Number of poles	6
Drive supplied by	Standard motor supplied by KSB - mounted by KSB	Motor bearing insulated	Yes
		Connection mode	Delta
Motor const. type	B3	Motor cooling method	Surface cooling
Motor size	315L	Motor material	Grey cast iron GG/CAST IRON
Efficiency class	IE2 acc. to IEC 60034-30	Frequency inverter operation allowed	FI allowed
Speed control selection	Speed adjustment		
Frequency invertor	Yes		

Materials C1

Volute casing (102)	Stainless steel 1.4408	Joint ring (411.10)	Thermoplastic PTFE-GF25
Casing cover (161)	Stainless steel 1.4408	Joint ring (411.31)	CrNi steel/graphite
Shaft (210)	Tempered steel C45+N	Seal cover (471)	Stainless steel 1.4571
Impeller (230)	Stainless steel 1.4408	Shaft protecting sleeve (524)	Stainless steel 1.4571
Bearing bracket lantern (344)	Grey cast iron JL 1040		

Certifications

Hydraulic performance test

Acceptance standard	ISO 9906 class 1 / 1B
Quantity meas. points Q-H	5
Certificate	Inspection cert. 3.1 to EN 10204
Test participation	Witnessed
Quantity, non-witnessed	0
Quantity, witnessed	1
NPSH test	Yes
Quantity meas. points NPSH	1
Vibration test	Yes

Hydrostatic test (room temp.)

Range	Complete pump without shaft seal
Test pressure	24.00 bar.g
Test time	10.0 min
Certificate	Inspection cert. 3.1 to EN 10204
Test participation	Non-witnessed

Material certificates: Volute casing, casing cover, impeller, shaft, seal cover, shaft protection sleeve (102, 161, 230, 210, 471, 524)

Certificate | Test report 2.2 to EN 10204

Material certificates: Bearing bracket lantern (344)

Certificate | Test report 2.2 to EN 10204

Material certificates: Impeller nut (922)

Certificate | Test report 2.2 to EN 10204

Performance curve

Customer item no.:
Order dated: 18/01/2013
Order no.: Anfrage TU HH
Quantity: 1

Number: ES 2149507
Item no.:300
Date: 18/01/2013
Page: 3 / 6

Version no.: 2

CPKN-C1 300-500
Chemical pump to EN 22858/ISO 2858/ISO 5199

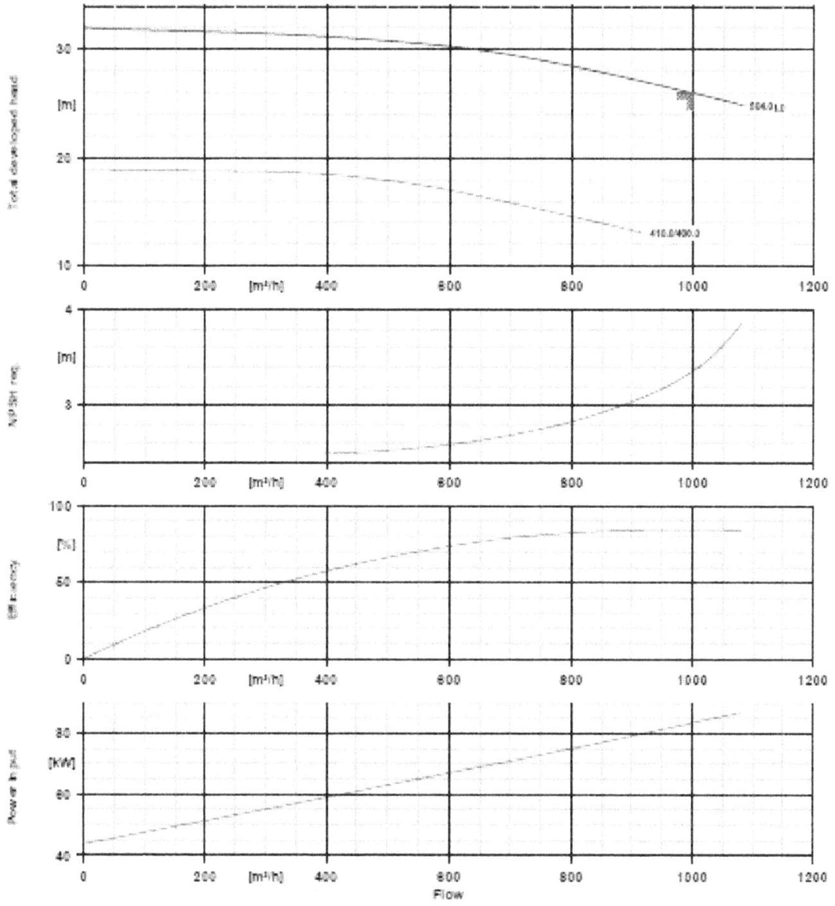

Curve data

Speed of rotation	889 rpm	Efficiency	84.7 %
Fluid density	998 kg/m³	Power absorbed	83.46 kW
Viscosity	1.00 mm²/s	NPSH required	3.38 m
Flow rate	1000.00 m³/h	Curve number	K2721.456/607
Requested flow rate	1000.00 m³/h	Effective impeller diameter	504.0 mm
Total developed head	26.00 m	Acceptance standard	ISO 9906 class 1 / 1B
Requested developed head	26.00 m		

Speed curve

Customer item no.:
Order dated: 18/01/2013
Order no.: Anfrage TU HH
Quantity: 1

Number: ES 2149507
Item no.:300
Date: 18/01/2013
Page: 4 / 6

CPKN-C1 300-500

Version no.: 2

Chemical pump to EN 22858/ISO 2858/ISO 5199

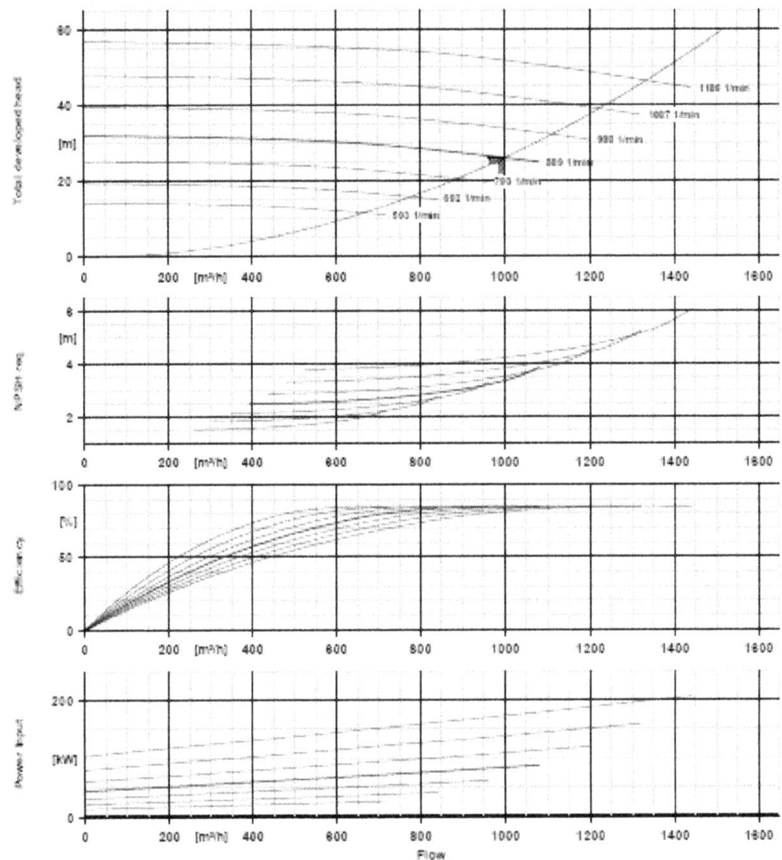

Curve data

Fluid density	998 kg/m³	Total developed head	26.00 m
Viscosity	1.00 mm²/s	Requested developed head	26.00 m
Requested flow rate	1000.00 m³/h	Effective impeller diameter	504.0 mm

Installation plan

Customer item no.:
Order dated: 18/01/2013
Order no.: Anfrage TU HH
Quantity: 1

Number: ES 2149507
Item no.:300
Date: 18/01/2013
Page: 5 / 6
Version no.: 2

CPKN-C1 300-500
Chemical pump to EN 22858/ISO 2858/ISO 5199

No drawing is available for the product as configured.

Drawing is not to scale

Dimensions in mm

Motor

Motor manufacturer	Siemens
Motor size	315L
Motor power	110.00 kW
Number of poles	6
Speed of rotation	988 rpm
Position of terminal box	0°/360° (top) Viewed from the drive

Baseplate

Design	Welded steel according to KSB AN 845
Size	17
Material	Steel ST
Leakage drain baseplate (8B)	Rp1. Without
Foundation bolts	M30x400

Connections

Suction nominal size DN1	DN 350 / EN 1092-1
Discharge nominal size DN2	DN 300 / EN 1092-1
Nominal pressure suct.	PN 16
Rated pressure disch.	PN 16

Coupling

Coupling manufacturer	
Coupling type	Metastream TSK
Coupling size	135
Spacer	200.0 mm

Weight net

Pump	1220 kg
Baseplate	780 kg
Coupling	
Coupling guard	
Motor	920 kg
Total	2920 kg

Connect pipes without stress or strain!

Dimensional tolerances for shaft axis height:	DIN 747
Dimensions without tolerances, middle tolerances to:	ISO 2768-m
Connection dimensions for pumps:	EN735
Dimensions without tolerances - welded parts:	ISO 13920-B
Dimensions without tolerances - gray cast iron parts:	ISO 8062-CT9

For auxiliary connections see separate drawing.

Connection plan

Customer item no.:
Order dated: 18/01/2013
Order no.: Anfrage TU HH
Quantity: 1

Number: ES 2149507
Item no.:300
Date: 18/01/2013
Page: 6 / 6

Version no.: 2

CPKN-C1 300-500
Chemical pump to EN 22858/ISO 2858/ISO 5199

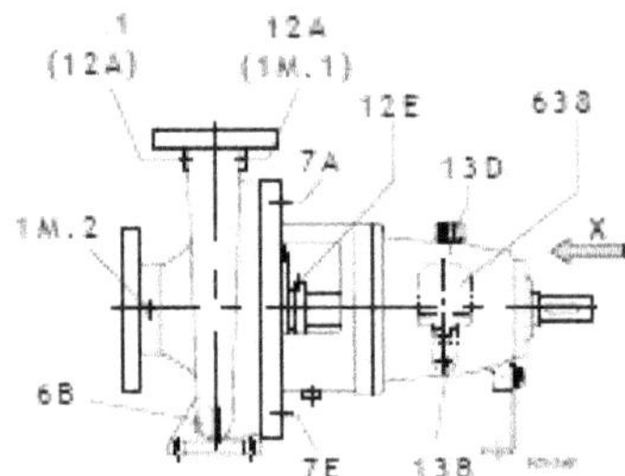

Connections

1M.1 Pressure gauge connection	G 1/2	Not executed
1M.2 Pressure gauge connection	G 1/2	Not executed
6B Pumped liquid drain	G 1	Drilled and plugged.
7E/7A Cooling liquid in/out	G 1/2	Drilled and plugged.
12E/12A Circulation in/out	G 1/4	Circulation line mounted by KSB
13B Oil drain	G 1/2	Drilled and plugged.
13D Refill / venting	Dia. 20	Closed with venting plug
638 Constant level oiler	Rp 1/4	Supplied unassembled with main equipment, to be installed by customer in acc. with operating instructions

Lebenslauf

Name	Szeliga
Vorname	Nicolai Sebastian
Staatsangehörigkeit	deutsch
Geburtsdatum	11.09.1987
Geburtsort	Reinbek
Geburtsland	Deutschland

09/1994 - 06/1997	Helmut-Landt-Grundschule Oststeinbek
08/1997 - 06/2006	Gymnasium Glinde
07/2006 - 04/2007	Zivildienst beim Deutschen Roten Kreuz e.V. in Glinde
10/2007 - 03/2012	Studium der Verfahrenstechnik an der Technischen Universität Hamburg-Harburg, Abschluss: Bachelor of Science
04/2012 - 10/2014	Studium der Verfahrenstechnik an der Technischen Universität Hamburg-Harburg, Abschluss: Master of Science
11/2014 - 03/2019	Wissenschaftlicher Mitarbeiter am Institut für Mehrphasenströmungen der Technischen Universität Hamburg, vom 01.04.2018 zudem in der Funktion als Gruppenleiter der Arbeitsgruppe „Industrielle Mehrphasenströmung“
04/2019 - 09/2019	Anfertigung der Dissertation am Institut für Mehrphasenströmungen der Technischen Universität Hamburg
seit 10/2019	Projekt-Ingenieur bei der Siemens AG in Frankfurt am Main

www.ingramcontent.com/pod-product-compliance
Ingram Content Group UK Ltd.
Pitfield, Milton Keynes, MK11 3LW, UK
UKHW061827190726
13853UKWH00009B/2471